GENE TECHNOLOGY

The INTRODUCTION TO BIOTECHNIQUES series

Editors:

J.M. Graham Merseyside Innovation Centre, 131 Mount Pleasant, Liverpool L3 5TF

D. Billington School of Biomolecular Sciences, Liverpool John Moores University, Byrom Street, Liverpool L3 3AF

Series adviser:

P.M. Gilmartin Centre for Plant Biochemistry and Biotechnology, University of Leeds, Leeds LS2 9JT

CENTRIFUGATION
RADIOISOTOPES
LIGHT MICROSCOPY
ANIMAL CELL CULTURE
GEL ELECTROPHORESIS: PROTEINS
PCR
MICROBIAL CULTURE
ANTIBODY TECHNOLOGY
GENE TECHNOLOGY

Forthcoming titles

LIPID ANALYSIS
GEL ELECTROPHORESIS: NUCLEIC ACIDS
PLANT CELL CULTURE
LIGHT SPECTROSCOPY
MEMBRANE ANALYSIS

GENE TECHNOLOGY

M.T. Dawson
National Diagnostics Centre, University College, Galway, Ireland

R. Powell
Department of Microbiology, University College, Galway, Ireland

F. Gannon
European Molecular Biology Organization, Postfach 1022 40, 6900
Heidelberg 1, Germany

βIOS
SCIENTIFIC
PUBLISHERS

© BIOS Scientific Publishers Limited, 1996

First published 1996

A CIP catalogue record for this book is available from the British Library.

ISBN 1 872748 76 7

BIOS Scientific Publishers Ltd
9 Newtec Place, Magdalen Road, Oxford OX4 1RE, UK
Tel. +44 (0) 1865 726286. Fax +44 (0) 1865 246823

DISTRIBUTORS

Australia and New Zealand
    DA Information Services
    648 Whitehorse Road, Mitcham
    Victoria 3132

India
    Viva Books Private Limited
    4346/4C Ansari Road
    Daryaganj
    New Delhi 110002

Singapore and South East Asia
    Toppan Company (S) PTE Ltd
    38 Liu Fang Road, Jurong
    Singapore 2262

USA and Canada
    Books International Inc.
    PO Box 605, Herndon, VA 22070

Typeset by Chandos Electronic Publishing, Stanton Harcourt, UK.
Printed by Information Press Ltd, Oxford, UK.

Contents

Abbreviations

ars	autonomously replicating systems
BAP	bacterial alkaline phosphatase
CAT	chloramphenicol acetyl transferase
CCC	covalently closed circular
CIP	calf intestinal phosphatase
DEPC	diethyl pyrocarbonate
DMSO	dimethyl sulfoxide
MCS	multiple cloning site
NMR	nuclear magnetic resonance
PCR	polymerase chain reaction
PNK	T4 polynucleotide kinase
PFGE	pulse field gel electrophoresis
RTase	reverse transcriptase
SDS	sodium dodecyl sulfate
Tdt	terminal deoxynucleotidyl transferase
UV	ultraviolet
VRC	vandyl ribonucleoside complex
YAC	yeast artificial chromosome

Preface

The genetic engineering revolution began over 20 years ago. It has evolved to the stage that we now have a universally used and greatly advanced technology, hence the title 'Gene Technology'. In Chapter 1, we attempt to mix history with information to give an appreciation of how the technology developed. The tools of gene manipulation are naturally occurring molecules termed enzymes; these are the subject of Chapter 2. In Chapters 3 and 4, we discuss the mechanisms and vehicles of gene manipulation, that is how DNA can be moved and prepared for analysis. In Chapter 5, we discuss how genetically engineered DNA is introduced into a host organism and perpetuated for subsequent analysis. In Chapter 6, we describe the methods for screening clones for a desired sequence. In Chapter 7, we discuss the experimental techniques required for the analysis and characterization of cloned sequences. In Chapter 8, we give a brief introduction to the polymerase chain reaction (PCR) which has in recent years revolutionized scientists' ability to study genes of interest. Finally, in Chapter 9, we discuss the implications of the technology in a broader context.

M.T. Dawson
R. Powell
F. Gannon

1 The Advent of Gene Technology

1.1 Development of the technology

Genetic engineering or the use of recombinant DNA technology has become so much a part of every aspect of modern biological research that it is easy to forget that it is a very recent addition to our scientific capabilities. To appreciate the importance of the developments in this area, it is appropriate to think back to the 1960s and to imagine oneself as a scientist interested in genes. It had been established by a variety of classical experiments that nucleic acid is the primary genetic material. In addition, it had been shown that the flow of information generally was from DNA to RNA to proteins. However, there are exceptions and it was subsequently shown in studies on retroviruses that the flow of information could occasionally be from RNA to DNA. As DNA was the fundamental molecule of inheritance it was inevitable that there would be a lot of interest in its structure and the manner in which it transmitted its information from generation to generation. Extensive work in this area by many groups using a variety of experimental systems led to a biophysical description of the structure of DNA which culminated in the definition of the double helix model of DNA by Watson and Crick in 1953 [1]. The intellectually satisfying unraveling of the structure of DNA and the clear description of how information was passed on from cell to cell during replication, left many scientists eager for a more thorough analysis of the genetic material.

The way appeared clear for the immediate study and description of the functional aspects of DNA; how genes are expressed in a specific manner; how DNA controlled cell cycles, and how it was involved in differentiation and development. The 'blueprint for life' had been

1

revealed and it seemed as if the task of building a monument of knowledge based on it would be a straightforward matter. However, scientists were to be frustrated for many years from achieving this goal. While biochemists could use their skills to isolate proteins in order to study their structure and characterize their primary amino acid sequence, those who were working on nucleic acids had to follow the advice of Shakespeare: "by indirections find we directions out". Wonderful intellectual games were played as the geneticists defined model systems that allowed them to generate mutations in the DNA sequence which they followed in biological assays of sometimes great complexity. The logic of these experiments was frequently close to that required of the mathematician. A number of parameters were established based on a hypothesis, a variant was introduced to the frequently artificial system and the outcome was then analyzed in the light of the parameters that had originally been defined. The intellectual satisfaction for the geneticists was immense. They did in fact succeed in defining the underlying rules of gene expression, replication, repair, etc. Progress was slow, however, and was frequently limited to specialist systems in prokaryotes, or simple eukaryotes such as yeast and fungi. But, as was shown later when gene structure in higher organisms was well understood, the reality was that the earlier experiments were heroic and of enduring importance but were far from universal in the truths that they revealed.

In the pre-genetic engineering world the underlying problem was that genes could not be readily isolated. This was inevitable when one considers the chemistry of DNA. DNA is relatively simple at the chemical level. In the broadest context DNA can be considered as a long featureless molecule. It is a polymer of several million units made up of only four basic building blocks called nucleotides (*Figure 1.1*).

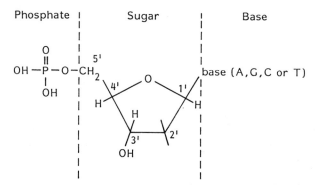

FIGURE 1.1: *Schematic diagram of a deoxynucleotide monomer. Note the position of the phosphate at the 5' position of the deoxyribose sugar. The 5' positionof this phosphate will be alluded to further in the text.*

Nucleotides have three components:

(i)   a pentose sugar (deoxyribose in DNA);
(ii)   one of four heterocyclic bases — adenine, guanine, cytosine and thymine; and
(iii) phosphoric acid.

The biologists' short hand for the nucleotides is A, G, C and T, which define the different bases (*Figure 1.1*). The four nucleotides are linked together to form a sugar phosphate backbone (*Figure 1.2*). How does the double helix form? One of the observations that led to the elucidation of the double helix structure was that the amount of A was equal to T and the amount of G was equal to C. From this followed the idea that A paired with T and G paired with C and this was how the two strands were attracted to each other. This model made absolute chemical and biological sense when the strands were positioned in an anti-parallel fashion (*Figure 1.3*). Thus the double helix is formed by the pairing of bases from complementary strands running in opposite directions [2].

Genetic studies had demonstrated that the genes were linked together in one continuous strand. Yet, there were no physical highlights which would indicate the start and finish of a gene. It was like a tape which when played contains a series of melodies but which when viewed by optical means has no distinguishing features. Without specialist equipment it is impossible to ascertain which part of the tape contains the musical passage of interest. Similarly, it was impossible to know which part of the DNA sequence coded for a gene.

Given these technical difficulties one could imagine that an expert group might have been formed to find the solutions. Such an expert committee hypothetically established in the 1960s might have made the following general recommendations to unblock the impasse and to obtain specific information on genes:

(i)   given the length of DNA, it must be broken up into smaller pieces to allow analyses to be performed;
(ii)   methods must be developed that will allow DNA fragments of similar physico-chemical characteristics to be separated;
(iii) an approach must be developed that will allow individual DNA fragments to be prepared in sufficiently large amounts for subsequent analyses.

Obviously no such committee was set up and the recommendations were not made. The concept that research can be directed in a knowing way is in fact challenged by such a failure. The reality was that

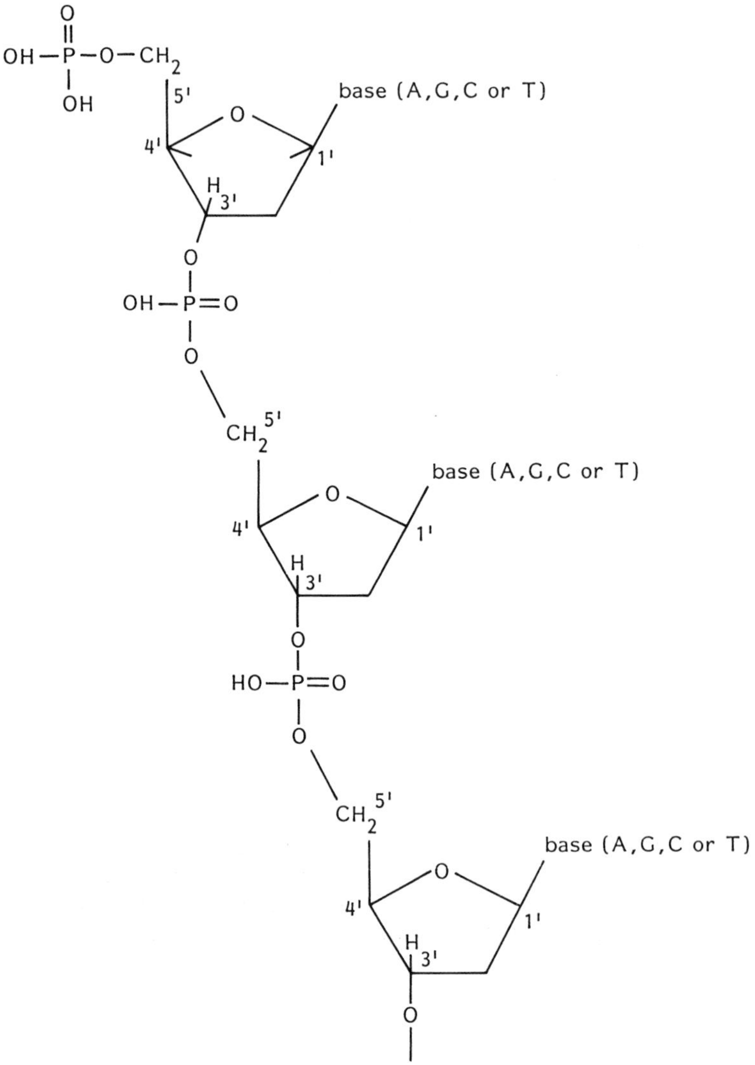

FIGURE 1.2: *Structure of nucleic acids. This schematic diagram shows how the nucleotide monomers are covalently linked in a sugar phosphate backbone. Note the linkages are at the 5′ and 3′ positions of the sugar. Also note that the bases are not involved in the backbone structure.*

many scientists working on individual projects for their own esoteric reasons put together a combination of technical tools which ultimately served as a mechanism to achieve these goals.

The three steps required for the isolation of a specific DNA fragment are considered below in more detail.

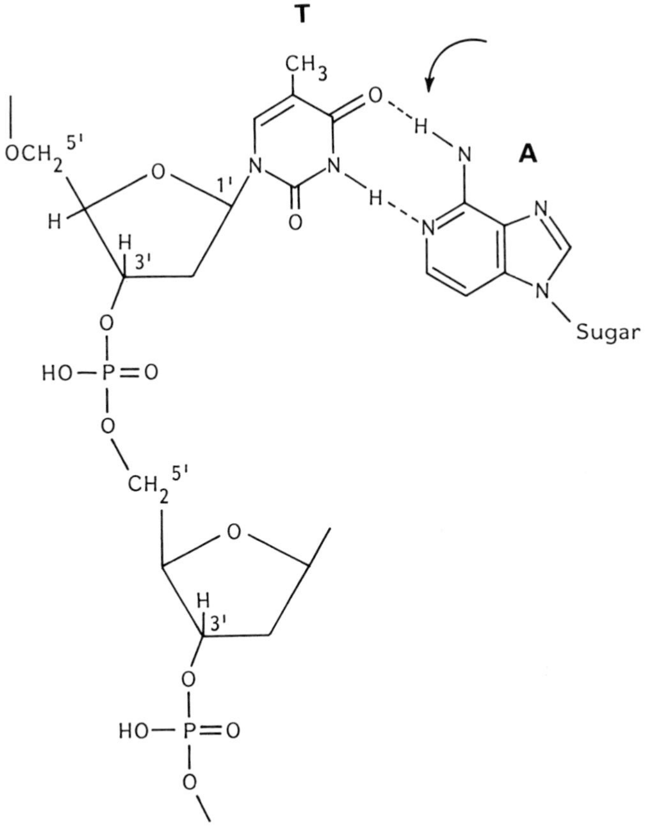

FIGURE 1.3: *Base pairing. This schematic diagram shows the manner in which base pairing occurs in an A–T base pair. Two hydrogen bonds provide the attraction between the bases. In the case of a G–C pair, three hydrogen bonds are involved. The hydrogen bonding is represented by a broken line (- - - - -).*

1.2 Isolation of specific DNA fragments

1.2.1 Fragmentation of the DNA

Because of its length, DNA can be sheared and broken down into smaller sizes by passing it through a narrow orifice (e.g. a syringe needle). An alternative physical method is ultrasonication. In this case the length of time that the DNA is exposed to the ultrasonic waves influences the length of the fragment. The fragment could be up to some several thousand bases in length which would be appropriate for many subsequent manipulations. However, physical approaches to fragmenting DNA never really took into consideration the subsequent steps that are necessary for the isolation of a gene of interest. One reason for this was that the fragments were random and therefore difficult to work with, another reason may have been the boundaries between those who were involved in the physical aspects of DNA research and those who were accustomed to a more biochemical approach. Biochemical approaches had been available for fragmenting DNA for many years. Enzymes called DNases (see Chapter 2) can digest DNA extensively. By carrying out incomplete digestions with these nonspecific DNases one could produce DNA fragments of varying length. However, these enzymes were difficult to control, not very specific and the physical methods were not used profitably to fragment the DNA prior to further detailed study.

A self-limiting enzymatic reaction that would yield gene-size fragments was needed. Such a major breakthrough came from studies on the mechanism by which bacteria protect themselves from invasion by foreign pieces of DNA. These studies showed that micro-organisms have a specific nuclease which scans incoming DNA for a particular sequence of bases and then cuts (digests) the incoming DNA at that sequence only (for details see Chapter 2). They subsequently became known as restriction enzymes. Endogenous DNA is protected either by not having that sequence or by a methylation event that modifies the target site for the nuclease thus rendering it resistant to cleavage. This was termed the restriction–modification system. One of the outcomes of these studies was the knowledge of the DNA sequences that were targeted by these enzymes. Target sites for the restriction enzymes generally can be from 4 to 8 base pairs (bp) in length. The recognition sequences for these enzymes are specific for a given organism and vary from organism to organism. It follows that in nature there are thousands of these restriction enzymes and in subsequent years several hundred of these have become readily available to scientists. As the restriction enzymes digest only at a given target site (under normal conditions) they can be left in contact with the DNA for a long period without yielding products different from those that would be anticipated from the DNA sequence.

The average size of the fragments that are generated is also implicit in the size of the target region. For example, consider a restriction enzyme that digests at a 6 bp target site. The theoretical chance of this site occurring is 1 in 4056 bases.. (There is a 1/4 chance of each nucleotide being present at any one site, and it follows that for all six the chance of them being present is, $1/4 \times 1/4 \times 1/4 \times 1/4 \times 1/4 \times 1/4 = 1/4056$). This is a size which would be useful if one wished to isolate what was thought to be a standard size gene. The molecular weight of an average protein can be considered to be approximately 50 000 Da and the average molecular weight of an amino acid is approximately 100 Da. It follows, therefore, that the average protein contains 500 amino acids. Given that the genetic code is triplet in nature this suggests that the average gene ought to be encoded by 1500 bases. Given the presumptions involved in this calculation one can see that the digestion of total DNA by a restriction enzyme, which views targets of 6-bp fragments, should yield many fragments that are about the size that might be expected to be necessary to accommodate a gene.

The mechanism of action of restriction enzymes presented a further major advantage. It had been shown by those working on restriction enzymes that the target site was not just a random (though specific) site. Rather it carried an internal symmetry such as that shown in the sequence recognized by the restriction enzyme *Eco*RI (see Chapter 2) and digests in an asymmetric manner. This means that the DNA, when digested, leaves some of the bases protruding from the target site. Because the protruding termini are in a single-stranded form it follows that they can associate with DNA fragments that are complementary to them. At one level it meant that restriction enzyme digests could be reversed. When considered in greater detail it is clear that not only can the digestion be reversed but two different fragments of DNA digested with the same restriction enzyme can in fact be linked to each other (see *Figure 2.2*). This inherent tendency of DNA digested by a restriction enzyme to bind to an unrelated fragment of DNA is at the core of recombinant DNA technology. This process, called ligation, required another enzyme which happily had become available and is known as DNA ligase.

1.2.2 Separation of DNA fragments

Given that the human genome contains over 3×10^9 bp and given the frequency with which a restriction enzyme directed against a 6-bp target site would digest, one can anticipate that the total number of different DNA fragments to be generated would be of the order of 10^6. Clearly it is not a trivial matter to isolate from 10^6 fragments of equivalent biophysical properties the one fragment that is of interest. Even today, using the most advanced separation technology, it is unlikely that such an experiment could succeed.

This is where the cross-fertilization of different scientific disciplines comes into play. If the biochemist provided the enzyme to digest the DNA, then microbiologists had in their laboratories trivialized the problem of separating identical entities. For years students and those who had no prior skills were able to take an overnight bacterial culture and spread it on an agar plate to produce separated colonies each of which arose from a single cell. This spatial separation of identical entities became a key factor in solving the problem of separating DNA fragments. Methods were devised that allowed the introduction of individual DNA fragments into micro-organisms. How to transfer each of the many fragments of DNA into a separate micro-organism? Fortuitously this did not present a major technical difficulty as over many years geneticists had mastered the skill of transferring DNA into micro-organisms. It was to them a core technology, and is described in greater detail in Chapter 3. So it was possible to envisage a separation method based on microbiological techniques allied to genetic skills for gene transfer and the biochemical method used to digest DNA.

1.2.3 Overcoming the problem of how to isolate large quantities of the gene of interest

The final requirement in the terms of reference put to the mythical committee was that the DNA fragment of interest should be made available in large quantities for further analyses. It was not a major problem to digest very large amounts of DNA at the start of the experiment, and indeed the process of transferring DNA fragments into a micro-organism does not have any real quantitative limitations either. However, it was soon realized by those trying to establish recombinant DNA techniques that simply transferring the DNA into the micro-organism was a futile exercise. The fragment of DNA that was transferred to an organism was very quickly diluted out upon growth and replication of the organism and lost. It became clear, therefore, that not only must a method for the separation of DNA be developed but also a method for propagating and amplifying the DNA that is transferred to the organism. The question then could be asked: how can the DNA that included the gene of interest be perpetuated in a micro-organism? Again the geneticists had the answers in their laboratories. For a variety of different motives many microbiologists and geneticists had been studying simplified systems which included origins of replication. Some of these studies were initiated to solve a medical problem, that of resistance to antibiotics. These studies had uncovered small independently replicating entities of DNA that carried on them the information required to inactivate antibiotics. Microbiologists working on the question of 'sexual' transmission of genetic material in bacteria had discovered similar extrachromosomal fragments, although they did not necessarily carry antibiotic resistance. These small extrachromosomal entities were termed episomes or R factors but with time have been grouped under the common name of plasmids. Many of these plasmids exist in multiple copies in a single bacterial cell.

Plasmids therefore represented one solution to the problem of how to replicate DNA once it entered the cell. If the DNA was linked to a plasmid then it could be perpetuated in multiple copies in the cell as the plasmid replicated in an autonomous manner. Furthermore, methods required for the transfer of plasmids to host bacteria were very well understood by microbiologists and geneticists. In addition, the cells that had successfully taken up a plasmid could be identified on a simple plate test by virtue of the antibiotic resistances that were carried on some of the plasmids.

Plasmids were not the only solution available when attempting to link the DNA fragments of interest to a replicating unit. Bacteriophages (bacterial viruses) which had become useful tools for geneticists in their study of gene transfer provided another option. The bacteriophage (phage) that was studied in most detail was lambda (λ) whose host is the bacterium *Escherichia coli*. Lambda phage can undergo an infectious life cycle (lytic) or can be integrated (lysogenic) into the *E. coli* chromosome. When it emerges from the chromosome in response to environmental conditions, it can carry with it a fragment of DNA that originates from the chromosome. Nature had shown, therefore, that this phage could act as a carrier of extra DNA, as would be required in a gene isolation experiment. Although each vector type had its advantages and disadvantages, the essential fact was that both bacteriophages and plasmids provided solutions to the problem of maintenance of the fragments of DNA that were of interest once they were transferred into the micro-organism. The mechanisms designed for the introduction of plasmids and bacteriophages into cells are covered in detail in Chapter 5.

These DNA elements soon became known by the common name of vectors, indicating their ability to carry target DNA into the micro-organism. However, not only did they succeed in that task but, because they are usually present in micro-organisms in very high copy numbers, the target DNA was present in each organism at a level of up to several hundred copies. The fact that the vector was transferred to a micro-organism also meant that not only were the DNA fragments separated and generated in large amounts but they could be grown in unlimited quantities subsequent to the identification of the micro-organism of interest. In this way the genetic engineers provided a solution to all the problems associated with the isolation of DNA fragments in sufficient quantities for future studies.

Almost all aspects of the scheme that the hypothetical committee might have wished to put together to isolate genes were therefore in place. The DNA could be digested into manageable fragments, a variety of vectors and methods to carry out the transfer of the DNA fragments to micro-organisms were available, and the methods to link the fragments to the vectors were available by digestion of the vector and the target DNA with the same restriction enzyme which, as will be shown in Chapter 2, generates compatible ends. Linked target and vector DNA could be introduced into host cells and propagated for further study (see Chapter 5).

A further advantage of this combination of methods was that at many steps the DNA was available for alteration or manipulation by the scientists. Again, a range of enzymes that allowed deletions and modifications of the DNA became available. Parts of the DNA could be removed, bases could be altered and new combinations of DNA could be put together. This added possibility of *in vitro* manipulation of the DNA became particularly important when some functional aspect of the DNA required detailed study or when the inserted DNA fragments were to be used primarily as a source of novel proteins from these micro-organisms.

Finally although genetic engineering is less than 20 years old we are already entering a new era. Previously a combination of biochemistry, genetics and microbiology was used to achieve the goal of isolating large amounts of DNA fragments of interest. Genetic engineering is now moving away from the use of micro-organisms to the use of enzymes to achieve this goal The prototype of these methods, the polymerase chain reaction (PCR) is described in Chapter 8. This technology has had a major impact on many aspects of science and has consolidated and facilitated the growth in the use of DNA methods by laboratories active in all basic and applied biological research. The importance of this development is such that a volume in this series is dedicated to PCR [3].

References

1.   Watson, J.D. and Crick, F.H.C. (1953) *Nature,* **171,** 946.
2.   Crick, F.H.C. and Watson, J.D. (1954) *Proc. Roy. Soc.,* **223,** 80.
3.   Newton, C.R. and Graham, A. (1994) *PCR.* BIOS Scientific Publishers, Oxford.

2 Manipulation and Analysis of DNA

2.1 Introduction

The study of nucleic acids by the application of genetic engineering techniques has two experimental prerequisites. The first is the ability to isolate intact DNA or RNA of sufficient purity to allow subsequent manipulation. The second is the ability to manipulate the isolated nucleic acids to produce the desired result. The major tools for manipulation are enzymes that catalyze specific reactions on nucleic acids. This chapter describes the various enzymes one can use and also the relatively simple laboratory methods one utilizes to monitor and analyze their activity.

2.2 Enzymes that modify nucleic acids

2.2.1 Restriction endonucleases

The ability to cleave or digest DNA in a specific and controlled manner is one of the fundamental procedures of genetic engineering. As indicated in Chapter 1, it was probably the discovery of enzymes that had the ability to cleave DNA at defined nucleotide sequences and methods for their purification that allowed gene technology to develop. These enzymes are termed restriction endonucleases or, more commonly, restriction enzymes. The discovery and characterization of bacterial restriction endonucleases led to the award of Nobel prizes to W. Arber, H. Smith and D. Nathans in 1978. Simply explained, these enzymes perform a precise and reproducible digestion of any DNA molecule if that molecule contains the specific sequence of nucleotides recognized by a restriction enzyme. This allows one to cleave DNA into defined fragments, often producing suitable quantities of DNA fragments for cloning experiments.

The DNA fragment termini produced by the action of these enzymes aid the joining of DNA fragments to form new recombinant DNA molecules. Also, a physical map can be obtained for a DNA molecule by the comparison of DNA fragment lengths produced on digestion with a series of these enzymes — termed restriction mapping. Given the length and apparently monotonous nature of DNA, it is important that we have markers to allow us to recognize the part of the DNA molecule on which we are working. An analogy would be a train line with no stations. While knowing where both ends are, you may be unsure of your exact location while traveling. However, the presence of stations along the route makes it far easier to describe the different regions of your journey and their characteristics. As restriction enzymes digest at specific sequences which occur infrequently along any fragment of DNA, it becomes convenient to describe any DNA fragment in terms of the location of specific restriction enzyme sites. Making such a physical map is an exercise in logic in which the locations of restriction enzyme sites along a linear stretch of DNA are deduced by a combination of digestions using a multiplicity of restriction enzymes. Finally, the importance of these enzymes to genetic engineering is seen by the fact that the choice of restriction enzymes is usually the first critical step in strategies to clone genes.

Given the importance of these enzymes and their central role in recombinant DNA technology, one could say that scientists and society were lucky that nature has 'invented' them in the first place. Their real function is as part of the defense systems of bacteria. They are used in a process called host-controlled restriction, to protect bacterial cells from infection by bacteriophage or other foreign DNA. In an act of wonderfully synchronized simplicity, the restriction enzymes digest the foreign DNA upon its entry into the bacterial cell. The bacterium's own DNA is protected from the action of its own restriction enzymes as its DNA is chemically modified; it contains extra methyl groups bonded to nucleotides thereby protecting them from the enzymes. Several hundred of these enzymes have been characterized from a large variety of bacterial species. They are classified as Type I, II and III, and it is Type II restriction endonucleases that are commonly used in genetic engineering.

A useful feature of Type II restriction endonucleases is that they will cleave DNA at a specific recognition sequence and nowhere else. For example, the enzyme *Eco*RI will cleave DNA only at the hexanucleotide sequence GAATTC while the enzyme *Bam*HI only recognizes the sequence GGATCC. The length of the recognition sequence can also vary. Examples of restriction enzymes which cleave at four-nucleotide sites are the enzyme *Sau*3A which recognizes GATC and *Alu*I which recognizes AGCT. It will have been noted that, in the preceding sentences, some strange words and acronyms are used to describe restriction enzymes. Such words have become the jargon of the trade and can, potentially, be a barrier towards understanding what is really being discussed. The key to entering this particular area of information is to understand that each restriction enzyme is given a name that is derived from the organism in which it occurs naturally. Occasionally these organisms have more than one restriction enzyme hence a number can be added. Thus *Eco*RI is a restriction enzyme from *E. coli* and *Bam*HI originates from *Bacillus amyloliquefaciens*. With so many enzymes available, it becomes pointless to recall their names and recognition site sequences. Happily the companies that sell these enzymes provide charts for easy access to this information.When giving the target sequence of a restriction enzyme, only one strand of the recognition site is described. When one considers both strands, one notices that almost all recognition sequences are palindromes, that is they are the same irrespective of the direction in which they are read (see *Figure 2.1*).

The next useful feature of the action of these enzymes for DNA manipulation is the nature of the termini of DNA fragments they produce. Some enzymes simply cleave DNA with a double-stranded break in the center of the recognition sequence. This produces DNA fragments with termini called blunt ends or flush ends (*Figure 2.1, Alu*I or *Hae*III).

The majority of the restriction enzymes produce DNA fragments with termini called cohesive ends or sticky ends. This is because the cleavage of the phosphodiester bonds of both DNA strands is asymmetrical or staggered, that is both strands are not cut at exactly the same position. Usually the stagger is over two or four nucleotides and this type of cleavage produces termini with short single-stranded overhangs (*Figure 2.1, Bam*HI or *Eco*RI).

This feature of the production of DNA fragments with defined ends (be they either blunt or sticky) is central to the ability to create new recombinant DNA molecules. If two DNA fragments have blunt ends, it is possible to join them together with enzymes called ligases (see Section 2.2.2). If two DNA fragments have sticky ends that are com-

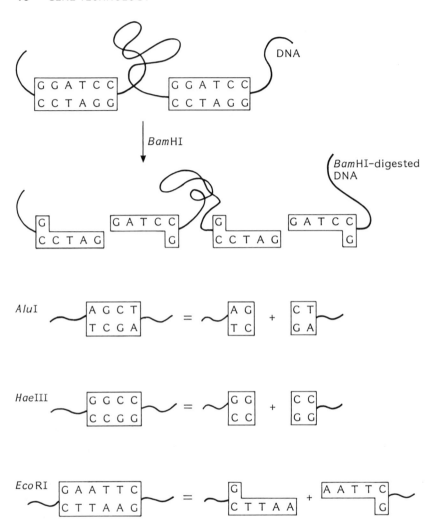

FIGURE 2.1: *Restriction endonucleases and their modes of action.*

plementary to one another, it is far easier to join them together with ligases as the sticky ends allow both molecules to join by hydrogen bonding between complementary bases before the ligases act. Conversely, if two DNA fragments have sticky ends that are non-complementary, the single strand regions must be removed (by making them double-stranded using polymerases or by the use of exonucleases) before they can be joined. The sole requirement for these ligation reactions is that the termini of the DNA fragments are compatible. It follows that fragments of DNA that are not adjacent to each other in a genome or which come from different genomes can be joined together in a recombinant DNA experiment.

Finally, where does one obtain these enzymes and how does one cleave DNA with them? Several hundred of these enzymes are now available, prepared to homogeneity by manufacturers of molecular biology products [1]. They are supplied in concentrated form with all the necessary information on the optimum conditions for their reaction with DNA. This includes details of optimum temperature, ionic requirements and amount of enzyme required to recognize and cleave DNA at a defined number of sites for a defined length of time. One then prepares a reaction mixture containing DNA, enzyme and buffered reagents followed by incubation for the desired time before analyzing the DNA cleavage, usually by gel electrophoresis (Section 2.3).

2.2.2 Ligases

The process of joining two DNA molecules together to form a new recombinant DNA molecule is catalyzed by a family of enzymes called ligases and the reaction is called a ligation. Ligases are enzymes found in all cells and they have a critical role in the replication and maintenance of nucleic acids. They catalyze the formation of the phosphodiester bond linking adjacent or closely juxtapositioned nucleotides. Owing to hydrogen bonding of complementary bases, double-stranded DNA molecules can retain their integrity if some of these bonds are missing, at least until a ligase arrives to seal the 'nick'. It is exactly this activity that allows ligases to be used in genetic engineering to join two DNA molecules. The basic requirement for all ligases is the presence of a minimum of one phosphate residue on the termini of the molecules of DNA which are to be brought together for ligation. There are two types of ligation reaction that can be performed depending on the ends of the fragments (*Figure 2.2*).

Sticky end or cohesive ligation (*Figure 2.2b*) is the more favorable method of joining two DNA molecules. The efficiency of ligation is increased by 50- to 100-fold over blunt-ended ligation. The only requirement is that the molecules to be joined share sticky ends that are complementary. This is ensured by digesting both DNA molecules with the same restriction enzyme. The complementary single-stranded regions at the ends of the molecules can base pair with one another, forming a relatively stable structure even if there are only two or four bases involved. This increases the length of time that two ends

(a) Blunt-end ligation

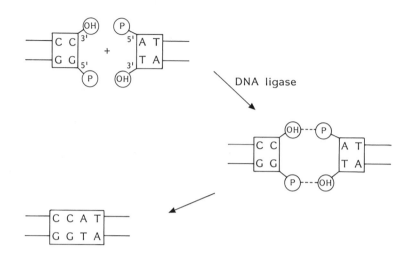

(b) Cohesive-end ligation

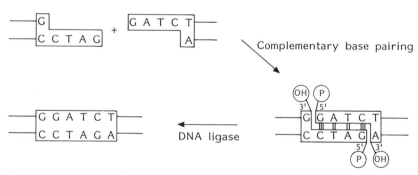

FIGURE 2.2: *DNA ligase and ligation reactions.*

are associated and therefore allows more time for the ligase to recognize and seal the nicks. By examining the sequences generated by restriction enzymes it becomes evident that different enzymes that recognize different sequences can result in complementary sticky ends. Therefore, DNA fragments produced by the action of such enzymes can be mixed and joined together by a ligase.

In blunt-ended ligation (*Figure 2.2a*), the formation of two phosphodiester bonds is catalyzed between the terminal bases of two DNA fragments with flush or blunt ends (i.e. no protruding nucleotides on either DNA strand). In practice, this is a very inefficient reaction as the ligase cannot pull two molecules together into close proximity to catalyze ligation. Therefore, high concentrations of DNA are required for blunt-ended ligation reactions. This increases the chances of two termini falling into close proximity. The opposite is true if one wishes to form a circular molecule by the blunt-ended ligation of the termini of a DNA fragment. In this case, high DNA concentrations would promote the joining of different DNA molecules (intermolecular ligation) forming dimers, trimers and eventual concatemers. A very low concentration might be inefficient but it would promote the chances of the two termini of one fragment coming into close proximity (intramolecular ligation) rather than termini of different molecules. In a strategic decision for a cloning experiment, the one advantage of blunt-ended ligation is the ability to join two DNA fragments, regardless of the sequence of their termini, provided they have blunt ends.

Like restriction enzymes, ligases may be obtained commercially. *E. coli* DNA ligase [2] and T4 DNA ligase [3] (obtained from *E. coli* cells infected with T4 bacteriophage) are the enzymes routinely used. T4 RNA ligase is also available and will catalyze the bonding of 5'-phosphoryl groups and 3'-hydroxyl groups of single-stranded DNA and RNA molecules.

2.2.3 Polymerases

Although the principal innovative step in genetic engineering experiments is the linkage of fragments of DNA from different sources, another important characteristic of this aspect of science is the fact that researchers can alter the DNA that is used in the experiment in a precise and selective manner. Occasionally this may mean the deletion of a part of the DNA fragment (using a restriction enzyme, for example) or the addition of an extra DNA fragment (again by the use of restriction enzyme digestions followed by the ligation of a new fragment). However, for some of the manipulations, the DNA may need to be modified and adjusted in a more delicate manner. It is common, in these circumstances, to use one or more of a wide variety of polymerase enzymes that have the general ability to add nucleotides to, or subtract nucleotides from, a pre-existing fragment of DNA. DNA polymerases are enzymes found in all cells and are responsible for DNA replication and maintenance. All DNA polymerases catalyze the addition of deoxyribonucleotides to the 3'-hydroxyl group of a terminal nucleotide on a primed, double-stranded DNA molecule using the pre-existing strand to direct the choice of nucleotide that is incorporated. An example of DNA polymerase activity is shown in *Figure 2.3*.

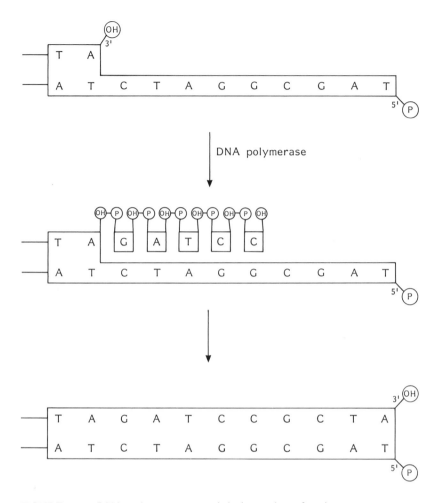

FIGURE 2.3: *DNA polymerases and their modes of action.*

Synthesis, therefore, is exclusively in a 5'–3' direction with respect to the newly synthesized strand. DNA polymerases can also have associated exonuclease activities. These exonuclease activities can be both 3'–5' and 5'–3' and are often useful in removing single-stranded termini from DNA fragments to produce blunt ends. There are a variety of DNA polymerases available, some suited to particular uses or experiments. The main uses of DNA polymerases in genetic engineering are the synthesis of DNA strands from a template, particularly when incorporation of a label such as a nucleotide linked to a radioisotope is required to prepare a DNA probe. They are also used in the synthesis of complementary DNA (cDNA) from mRNA for cDNA cloning.

As recombinant DNA techniques have diversified, scientists have defined, with greater specificity, the properties they require from different polymerases. In parallel, biochemical experiments characterizing these different polymerases have been carried out and these enzymes are now readily available to the scientific community. The differences between these enzymes frequently reside not in their polymerization ability but in associated exonuclease activities (i.e. DNA degradation) that they possess. For example, *E. coli* DNA polymerase I (Pol I) has 5'–3' exonuclease activity, whereas Klenow fragment, which is a portion of *E. coli* Pol I, has no 5'–3' exonuclease activity [4]. Both of these polymerases are commonly used in DNA labeling reactions to provide DNA probes using methods called nick-translation and oligo-labeling (see Chapter 6). Another example is T7 DNA polymerase, now commonly used for DNA sequencing.

These polymerases are all prepared from *E. coli* cells and so have an optimum temperature of 37°C; higher temperatures result in their denaturation. DNA polymerases from thermophilic organisms such as *Thermus aquaticus Taq* polymerase, with their ability to catalyze reactions at extreme temperatures have revolutionized many aspects of genetic engineering, especially the polymerase chain reaction (PCR) (see Chapter 8). Variations on the different uses of the DNA polymerases will be found as the chapters of this book unfold. However, there are some other polymerizing enzymes that should be introduced at this stage.

Terminal deoxynucleotidyltransferase or terminal transferase or, more simply, TdT, is a DNA polymerase that can catalyze the incorporation of deoxyribonucleotides to the 3'-hydroxyl group of a terminal nucleotide without the need for a primed template [5]. This reaction is often called 'homopolymer tailing'. Single-stranded DNA is the preferred substrate but double-stranded DNA can also be tailed, particularly if it has a 3' single-stranded terminus. As it requires no template, the nucleotides added to the 3' end depend on those present in the reaction. Therefore, one can produce a molecule with a series of guanine residues at the 3' ends by simply incubating the DNA in the presence of dGTP and terminal transferase (*Figure 2.4*). Homopolymer tailing was once the method of choice for joining two DNA molecules together (both molecules were tailed separately with complementary nucleotides and then mixed and allowed to join by base pairing; *Figure 2.4*) but it is not in routine use today. However, it still has a useful role in the specific labeling of DNA molecules at their 3' termini and in some PCR experiments.

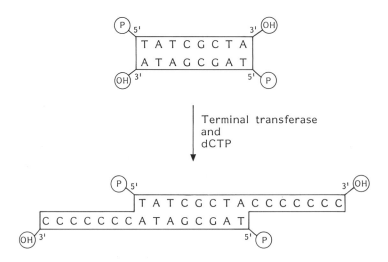

FIGURE 2.4:   *Terminal transferase and homopolymer tailing.*

Another enzyme that has played a major role in the development of genetic engineering is reverse transcriptase. This is an enzyme that uses an RNA template to produce complementary DNA (cDNA) in a 5'–3' direction. These enzymes were originally isolated from eukaryotic cells infected with RNA retroviruses. These viruses encode an enzyme that allows them to synthesize a DNA copy of their RNA genome which can then insert into the host genome. When first described, this enzyme was viewed as controversial in that it resulted in the flow of genetic information from RNA to DNA, contradictory to what had originally been thought. Happily, this enzyme became available to the scientific community at the beginning of the genetic engineering era. Like all the previously described enzymes, this natural activity has been extremely useful in genetic engineering. It is used for the cloning of RNA molecules, in particular eukaryotic mRNA molecules. As mRNA molecules are single-stranded, chemically labile and not readily integrated into vectors (see Chapters 3 and 4) the ability to convert them into the far more stable double-stranded DNA using reverse transcriptase at an early stage allows their subsequent treatment as DNA molecules [6]. *Figure 2.5* illustrates the mode of action of reverse transcriptase.

The range of polymerases which are used by scientists is ever increasing. Examples include phage RNA polymerases such as SP6, T7 and T3, which are becoming very widely used. All three enzymes recognize their own or specific promoters and transcribe RNA molecules from

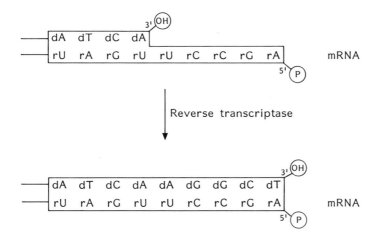

FIGURE 2.5: *Use of reverse transcriptase to prepare cDNA from mRNA.*

the promoters in large quantities. One application for these enzymes is to incorporate the promoters into gene vectors beside the site of insertion for foreign DNA. By use of the appropriate RNA polymerase, large quantities of RNA derived from the insert can be prepared rapidly. These can be useful as RNA probes if a radioactive ribonucleotide is included in the reaction. Also, the RNA transcripts can be translated *in vitro* to generate proteins for functional studies.

2.2.4 Methylases

Methylases are the family of enzymes previously described that form a partnership with the restriction endonucleases. They are the enzymes responsible for protecting the host bacterial DNA from degradation by the restriction enzymes. Each bacterium has its own methylase. *Eco*RI methylase, isolated from *E. coli*, is a methylase used specifically by genetic engineers to protect DNA molecules from the action of the restriction enzyme *Eco*RI. This may be important if one does not know if a DNA fragment to be cloned contains an internal recognition site for *Eco*RI. If *Eco*RI is the cloning enzyme and the DNA fragment has one or several recognition sites, then this fragment will be cleaved into one or more fragments by the action of the enzyme and some of the resulting DNA fragments could be lost. Therefore, it may prove impossible to find all of these fragments and 'restrict' them. The best option is to use *Eco*RI methylase to modify and protect any *Eco*RI sites. The usefulness of this procedure can be seen in the ligation of adaptors or linkers to cDNA molecules for cloning purposes (Chapter 4).

2.2.5 Kinases

T4 polynucleotide kinase (PNK) catalyzes the transfer of the terminal phosphate of ATP on to the 5'-hydroxyl group of a terminal nucleotide of DNA or RNA [7] (*Figure 2.6*). This reaction is most commonly used to label nucleic acids specifically at their 5' termini. Radioactive ATP (e.g. [γ-32P]ATP) is used as substrate and T4 PNK is used to add the terminal radioactive phosphate group on to the 5' terminus of the molecule to be labeled. This is the method of choice for labeling synthetic oligodeoxynucleotides for use as probes. It can also be used to aid the ligation of such sequences.

2.2.6 Phosphatases

Bacterial alkaline phosphatase (BAP) and calf intestinal phosphatase (CIP) are enzymes that catalyze the removal of 5'-phosphate groups from DNA and RNA (*Figure 2.7*).This activity can often prove useful in cloning experiments. Any DNA molecule that has been treated with either phosphatase enzyme has lost the ability to self-ligate. As there are no phosphate groups at the 5' termini, the ligases cannot catalyze a phosphodiester bond formation. Therefore, if one wishes to construct a recombinant DNA molecule from a mixture of two molecules, it is possible to ensure that one of the molecules can only ligate to the other by treating it with a phosphatase. This is particularly useful when constructing recombinant clones using plasmids as the vector to accept foreign DNA fragments. By treating a linearized plasmid with phosphatase, one can ensure that it cannot self-ligate and circularize. Therefore, a high background of clones representing cells with recircularized plasmid is avoided. Similarly, foreign DNA to be ligated into

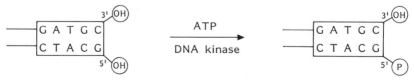

FIGURE 2.6: *Addition of phosphate groups to DNA by kinases.*

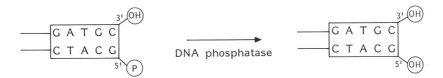

FIGURE 2.7:   *Removal of phosphate groups using a phosphatase.*

a vector can be phosphatased. This means that foreign DNA can ligate only to plasmid DNA molecules, thus avoiding isolating clones that contain noncontiguous fragments of DNA.

2.2.7 Exonucleases

Lambda exonuclease and exonuclease III (Exo III) are two DNA-degrading enzymes of particular use in some experimental situations. Unlike the restriction endonucleases, these exonucleases attack and degrade DNA from the ends. Lambda exonuclease catalyzes the degradation of double-stranded DNA in the 5'–3' direction and Exo III catalyzes the degradation of double-stranded DNA in the 3'–5' direction. The main use of these enzymes is to modify existing DNA fragment ends for deletion analysis.

2.3 Gel electrophoresis of nucleic acids

Along with the discovery, characterization and purification of enzymes for nucleic acid manipulation, the development of techniques to characterize DNA or RNA molecules was central to the advances in genetic engineering. Gel electrophoresis became the technique of choice and this was not very surprising as other biomolecules such as proteins were being analyzed by this technique. Nucleic acid molecules, like proteins, have an electric charge. For DNA and RNA it is always negative due to the oxygen atoms associated with the phosphate group joining nucleotides on both DNA strands. Therefore, if nucleic acid molecules are placed in an electric field, they will move or migrate towards the positive electrode (anode). However, due to the nature of nucleic acids, this migration has no analytical use unless the electrophoretic migration is performed through a gel as a support medium. These gels, normally agarose or low-concentration polyacrylamide, provide a matrix of pores through which the nucleic acids must migrate towards the positive electrode. The analytical principle of this procedure is that smaller nucleic acid molecules can migrate at higher speeds through the pores. Therefore, gel electrophoresis will separate DNA and RNA molecules according to their size [8]. Estimation of fragment lengths can be calculated using a formula correlating migration rate to molecular weight:

$$D = a - b(\log M)$$

where D is distance migrated, M is the molecular weight and a and b are constants determined by electrophoretic conditions. Routinely, however, DNA fragment lengths are estimated by electrophoresing or, more colloquially, 'running' known molecular weight size markers in parallel with the samples on gels to prepare a standard calibration curve correlating distance migrated with fragment length. The migration distance of an unknown fragment can then be placed on the curve and an estimation of its length calculated.

However, the migration distances of nucleic acids during electrophoresis depend not just on their lengths but also on the shapes of the molecules. Single-stranded nucleic acid molecules can form a variety of structures through intramolecular hydrogen bonding and more compact molecules migrate faster than might be expected from their molecular weight. In order to get around this problem one can use denaturing gels that contain denaturing agents such as urea (6–8 M), formamide (50–90%) or methyl mercuric hydroxide (5 mM) which ensure that the DNA remains as a linear molecule.

In order to see and analyze the results of a gel electrophoresis, one must have a method of staining the nucleic acid to make it visible. The chemical routinely used is ethidium bromide [9]. This chemical has two properties that make it useful for nucleic acid visualization. The first is that it binds to nucleic acid by slipping (intercalating) between neighboring bases of nucleotides. The second property is that it fluoresces in UV light when bound to DNA. Therefore, by simply staining a gel after electrophoresis in a dilute (1–5 µg ml⁻¹) solution of ethidium bromide for a few minutes, the migrated DNA will take up the stain and, if placed over a UV illuminator, bright pink bands representing populations of DNA fragments can be visualized and photographed.

An alternative method for visualization of nucleic acids after gel electrophoresis is to incorporate a radioactive marker into the nucleic acid for examination. This is often a nucleotide with a radioactive atom attached (e.g. [^{32}P]dNTP or [^{35}S]dNTP) and can be incorporated into a nucleic acid by one of the polymerase enzymes. The nucleic acid is then termed 'labeled' or 'hot' and, following electrophoresis, the gel may be exposed to X-ray film and later developed. The appearance of black bands on the film indicates the presence of the labeled nucleic acid. This autoradiographic method of visualizing nucleic acid is extremely sensitive when compared with ethidium bromide (1 pg versus 1 ng) but requires training in the use of radioisotopes. Care is required with both methods as both ethidium bromide and radioisotopes are powerful mutagens.

2.3.1 Agarose gel electrophoresis

Agarose gel electrophoresis is undoubtedly the routine technique for the analysis of nucleic acid molecules, particularly to monitor the success of nucleic acid isolation procedures and to analyze enzymatic manipulations such as restriction enzyme digestion. The gels are prepared by dissolving agarose powder in an electrolyte solution (electrophoresis buffer), boiling and casting it in a gel mould. After allowing the gel to cool to about 50°C a comb is placed across the end of the molten gel to form sample wells around which the agarose sets. After the gel has set, the comb is removed, leaving wells into which nucleic acid samples are loaded. The gel slab is removed and placed in a horizontal electrophoresis chamber, submerged in electrophoresis buffer and an electric charge is applied. Depending on the percentage agarose of the gel, DNA fragments ranging from 0.5 to 30 kb in length can be clearly resolved. Rapid analysis can be performed in 20–30 min with the longest separations requiring 8–12 h of electrophoresis. Nucleic acids are then visualized, usually by ethidium bromide staining. *Figure 2.8* shows a photograph of DNA fragments after agarose gel electrophoresis and ethidium bromide staining

2.3.2 Polyacrylamide gel electrophoresis (PAGE)

The rationale for polyacrylamide gel electrophoresis or PAGE is identical to that of agarose. However, the resolution power for defining DNA fragments of different lengths is far greater and within a size range where agarose gel electrophoresis is less informative. This is shown by the use of PAGE systems for elucidating the nucleotide sequence of DNA molecules. In this case, DNA fragments differing in length by only one base can be resolved accurately. PAGE systems are therefore the method of choice for resolving DNA fragments of 1–500

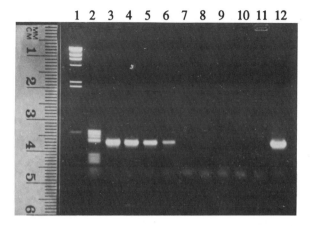

FIGURE 2.8: *Agarose gel electrophoresis of DNA fragments. Photograph of an agarose gel stained with ethidium bromide and visualized over an ultraviolet source. Lanes 1 and 2 contain a mixture of DNA fragments of known length. These can be used as reference points to determine the length of DNA fragments of interest. In this case, a DNA fragment of approximately 400 bp is seen in decreasing concentrations in lanes 3–6. No DNA fragments are apparent in lanes 7–11. Lane 12 represents the positive control for this analysis and the 400-bp DNA fragment can be seen.*

bp in length. The gels, usually containing urea, are cast between two glass plates (often with only a 1–2 mm space between them) and the gel system is vertical. Visualization of nucleic acids is by ethidium bromide staining, by autoradiography or by the use of lasers in automated systems. *Figure 2.9* shows a photograph of a DNA sequencing gel after PAGE and autoradiography.

2.3.3 Pulse field gel electrophoresis (PFGE)

This technique is a more recent development and is used to separate DNA fragments of chromosome size. The matrix is similar to that of agarose gel electrophoresis but the application of the electric current is different. The voltage is reversed between the poles at regular intervals of time (usually periods of seconds) during the electrophoresis. This causes the DNA fragments to 'turn around' and migrate in reverse as the voltage is switched [10]. However, very large DNA fragments of 100 kb–1 Mb in length cannot simply 'turn around' in the time periods involved with the resulting effect that these molecules

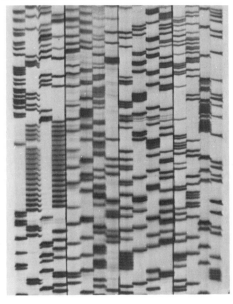

FIGURE 2.9:   *Polyacrylamide gel analysis of DNA fragments. Photograph of a portion of an autoradiograph or X-ray of a DNA sequencing gel. Each DNA clone (1–4) is represented by four lanes (G, A, T, C), each lane corresponding to a particular nucleotide. Each black band on the autoradiograph represents a DNA fragment. This type of gel electrophoresis has the resolving power to separate DNA fragments differing in length by only one base.*

migrate extremely slowly. This technique, along with similar variations, has been shown to resolve DNA molecules between 5 and 10 Mb in length. Therefore, chromosomes from simple eukaryotic organisms, such as yeast, can be resolved. As yet, human chromosomes that are between 50 and 400 Mb cannot be resolved by this technique, but extremely long DNA fragments generated by restriction enzymes with target sites that are very rare and underrepresented in DNA may be usefully analyzed in this manner.

2.4 Conclusions

In this chapter we have introduced some more of the reagents and methods that are central to the current techniques known collectively as genetic engineering. It is clear that the molecular biologist has an extensive arsenal of enzymes and methods that can be used to achieve a specific goal. Equally, these enzymes ensure that the limiting factor in experiments is rarely technical feasibility itself; each molecular biology freezer usually has a wide range of these reagents on hand for daily use. This contrasts starkly with the advent of genetic engineering when the skills of the biochemist in purifying the various enzymes were the major limitation. Now it is merely a matter of having sufficient time and money to follow the experimental paths that are plotted out.

References

1. Roberts, R.J. (1988) *Nucleic Acids Res.* (suppl.), **16**, r271.
2. Panasenko, S.M., Cameron, J.R., Davis, R.W. and Lehman, I.R. (1977) *J. Biol. Chem.*, **253**, 4590.
3. Weiss, B., Jacquemin-Sablon, A., Live, T.R., Fareed, G.C. and Richardson, C.C. (1968) *J. Biol. Chem.*, **243**, 4543.
4. Klenow, H. and Henningsen, I. (1970) *Proc. Natl Acad. Sci. USA*, **65**, 168.
5. Roychoudhury, R., Jay, E. and Wu, R. (1976) *Nucleic Acids Res.*, **3**, 101.
6. Berger, S.L., Wallace, D.M., Puskas, R.S. and Eschenfeldt, W.H. (1983) *Biochemistry*, **22**, 2365.
7. Richardson, C.C. (1971) in *Procedures in Nucleic Acid Research* (G.L. Cantoni and D.R. Davies, eds), Vol. 2. Harper and Row, New York, p. 218.
8. Helling, R.B., Goodman, H.M. and Boyer, H.W. (1974) *J. Virol.*, **14**, 1235.
9. Sharp, P.A., Sugden, B. and Sambrook, J. (1973) *Biochemistry*, **12**, 3055.
10. Carle, G.F., Frank, M. and Olson, M.V. (1986) *Science*, **232**, 65.

3 Generation of Genomic DNA Libraries

3.1 Introduction

DNA libraries (banks) need to be produced in order that a DNA sequence(s) or gene(s) of particular interest may be isolated from a multitude of unwanted sequences. In other words, it is equivalent to finding a needle in a haystack. In general this is achieved by isolating DNA from the source of the desired sequence (target DNA) followed by modification with appropriate enzymes and joining (see Chapter 2 and Section 3.2.3) with specialized DNA molecules (vector DNA). This is followed by the introduction of the new molecular combinations (recombinant DNA molecules) into a suitable host which allows their propagation on agar plates, either as discrete colonies or plaques (see Section 3.2.3). If a sufficient number of recombinant colonies or plaques are produced a DNA library will be generated.

The strategies involved in generating a library depend on the source of the target DNA. The size of the target organism's genome can vary enormously. For example, there are over 3×10^9 base pairs in the entire human genome as distinct from an average of 3 million base pairs in a bacterial genome. Therefore, the logistics and practicalities of making libraries from these organisms are obviously different and will be considered in subsequent chapters.

Target and vector DNA are the molecules that need to be joined together and thus are the starting points for the generation of any DNA library. DNA vectors are the vehicles by which target DNA can be introduced into a chosen host. There are four major categories of vector. These are plasmids, bacteriophages, cosmids and, for larger DNA sequences, yeast artificial chromosomes (YACs). The choice of vector is largely determined by the requirements of a given experiment. Therefore, as each vector type is described the context in which it is used will be explained. In Section 1.2.3 the background to how plasmids and bacteriophages became the 'chosen vectors' was covered. However, as genetic engineering has evolved a parallel sophistication in vector development has occurred. Therefore, it would be remiss not to consider in some detail the variety of generic DNA cloning vectors used in gene technology.

3.2 Vectors used in DNA cloning

3.2.1 Plasmid vectors

Plasmids are found in bacteria and are autonomously replicating extrachromosomal circular DNA molecules [1]. Those used as cloning vectors are specifically designed for the purpose of gene cloning and have certain minimal requirements. They are small, between 2 and 8 kb, and often have a high copy number. They also encode antibiotic resistance genes for propagation in a suitable host and contain unique restriction enzyme sites for cloning purposes (see *Figure 3.1*). Plasmids are frequently used as they are the easiest class of vector to work with from the point of view of experimental manipulation, that is isolation from bacteria, modification with enzymes, ligation and introduction into their hosts. The major disadvantage of plasmid vec-

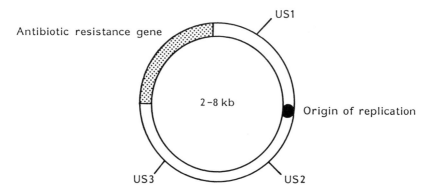

FIGURE 3.1: *A schematic representation of a plasmid vector showing the minimal requirements of an antibiotic resistance gene, an origin of replication and an arbitrary number of unique restriction sites designated US1–3.*

tors is that the size of the target DNA that can be inserted into these vectors is usually limited. In addition, recombinant plasmids can be extremely unstable in certain hosts, for example some strains of *Bacillus*.

The plasmids used in the pioneering gene cloning experiments fulfilled the aforementioned minimal requirements. One of the earliest vectors was pBR322 which was named after the scientists Bolivar and Rodriguez who were involved in engineering this plasmid (*Figure 3.2*). pBR322 is based on the ColE1 element [2] and is a small plasmid of 4.36 kb in size [3]. It replicates extrachromosomally in the bacterium *E. coli*. pBR322 encodes antibiotic resistance genes for both ampicillin (*Amp*) and tetracycline (*Tet*) which facilitate selection on antibiotic-containing plates. It also contains unique restriction sites for many enzymes which include *Sal*I, *Eco*RI, *Bam*HI, *Sal*I and *Pst*I. One feature of pBR322 that makes it useful from the point of view of selection and screening is the presence of the two antibiotic resistance genes. For example, restriction digestion of pBR322 with *Sal*I allows the insertion and ligation of *Sal*I-digested foreign DNA into the middle of the *Tet* resistance gene (*Figure 3.2*). This facilitates screening of transformed bacterial colonies for *Amp* resistance which will be intact following integration of target DNA. It also allows one to screen subsequently for *Tet* sensitivity which arises from the interruption of the *Tet* resistance gene in order to identify those containing recombinant DNA. This way of selecting for recombinants has now been largely replaced by phenotypic screening which allows direct visualization of

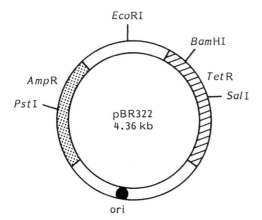

FIGURE 3.2:   *Schematic representation of the plasmid pBR322. This figure shows the* Amp *and* Tet *resistance genes (shaded regions of circle) and the origin of replication (ori). Some of the unique restriction enzyme sites are also indicated (*Bam*HI, Pst*I*, Eco*RI *and* Sal*I).*

recombinants on the plate as described in the following paragraphs. The earlier method of antibiotic sensitivity testing was tedious and time consuming as it involved replica plating of the individual colonies. In some cases this involved picking thousands of individual colonies.

As genetic engineering advanced and diversified, the pBR322-based plasmid vectors were modified to be more versatile and user friendly to the extent that the new derivatives now bear little resemblance to the generic pBR322 vector except at the level of their essential features. Plasmid vectors now available, either commercially or from research groups, come in a variety of sizes and with many specialized features, which make it easier to clone DNA. Individual vectors usually have a range of distinct, desirable features that have been engineered into the plasmid to make them more convenient to use.

One revolutionary feature which was introduced into the pUC family of plasmids was the presence of a multiple cloning site (MCS). The MCS is a short stretch of DNA which contains the recognition sequences for a large number of restriction enzymes, thereby increasing the choice of enzyme for digestion of the vector and target DNA samples. An MCS region is now an essential feature of most commercially available plasmid vectors. Another major development in vector improvement was the introduction of the *lacZ* gene (which codes for the protein β-galactosidase) into a number of plasmid vectors (*Figure 3.3*). When the *lacZ* gene is expressed in a suitable host it is capable of degrading the chromogenic substrate X-gal (5-bromo-4-chloro-3-indolyl β-D-galactopyranoside) to produce a blue color. If a DNA frag-

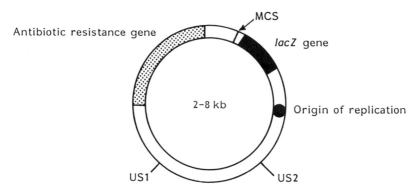

FIGURE 3.3: *A schematic representation of a generic plasmid vector showing the features described in* Figure 3.1. *In addition the gene for β-galactosidase (*lacZ*) is inserted into the vector. This can be used to screen for inserted DNA as described in the text. MCS, multiple cloning site.*

ment is introduced into the *lacZ* gene the β-galactosidase protein is not produced and hence any clone with an insert is incapable of degrading the X-gal substrate. Thus, recombinants are colorless and nonrecombinants are blue.

The use of plasmid vectors in the generation of DNA libraries is limited. This is mainly due to their inability to maintain faithfully and propagate large amounts of cloned DNA. In general, plasmids are limited in the size (<10 kb) of insert that can be conveniently cloned. Therefore, plasmid vectors are usually restricted to libraries of small genomes, for example bacteria, but are the universal choice for all manipulation of isolated genes that are subcloned for subsequent studies.

3.2.2 Bacteriophage vectors

In many instances bacteriophages are the favored vectors for generating DNA libraries, which is primarily due to the capacity of some bacteriophage vectors to accommodate large fragments of DNA. However, before discussing bacteriophage vectors in detail, we will consider the biology of bacteriophage and its relevance to the development of genetic engineering.

Bacteriophages are viruses that infect bacteria. They have a requirement for host functions for their propagation. A number of different bacteriophages have been studied and well characterized. However, the vast majority of gene cloning experiments are performed with bacteriophage lambda (phage λ) and derivatives thereof. Therefore, we will limit detailed discussion of bacteriophage vectors to this phage which infects *E. coli*. Infective phage λ exists external to its basic protein and DNA body. It has a protein head which contains the phage DNA (approximately 49 kb). It also has a tail and tail fibers (*Figure 3.4*). The tail fibers allow the phage to adsorb on to the surface of the bacterium (imagine Apollo 11 landing on the surface of the moon!). The phage tail contacts the surface of the bacterial cell and the phage DNA is injected internally. Once inside the bacterial cell the λ DNA undertakes either a lytic or lysogenic cycle. In the lytic cycle the cell's machinery is taken over by the λ DNA. The phage DNA is replicated, transcribed and translated following which phage assembly occurs. When a threshold number of phage are assembled, the cell will burst releasing the mature phage which can infect other bacterial cells. In the lysogenic cycle the phage DNA enters the cell and is integrated into the host chromosome. The decision whether to lysogenize or

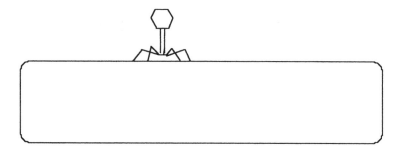

FIGURE 3.4:   *A schematic representation of a bacteriophage contacting the surface of a bacterial cell. Following adsoprtion, the phage DNA is injected into the bacterium.*

undergo the lytic cycle is complicated at the molecular level and is probably one of the most well studied genetic control mechanisms [4].

The biological properties of phage λ have made it an extremely useful tool for the generation of DNA libraries. Phage λ DNA can be modified in several ways that allow it to be used as a vector for cloning target DNA. Infection is achieved by packaging recombined DNA into phage outside the natural environment by a process known as *in vitro* packaging. *In vitro* packaging has been made possible by the isolation of a combination of specific phage mutants that can overproduce the various protein components of the mature phage while lacking one of the essential proteins. These can be mixed together in the correct proportions to generate packaging mixes. When concatemeric phage λ DNA is added to the packaging mix it is packaged by a specific cleavage mechanism at a sequence termed the cos site (*Figure 3.5*). Thus infective phage can be generated *in vitro*. These can then be used to infect *E. coli* cells. The infected cells are grown briefly and subsequently plated out on agar plates. Successful infection results in the production of small clearings on an agar plate which are termed plaques.

There are two broad classes of phage λ vector used in gene cloning. These are termed replacement and insertion vectors. Replacement vectors have been designed to accommodate large fragments of DNA. This is achieved by removal from the viral genome of DNA which is unnecessary for its function by restriction digestion. The removed fragment is termed the 'stuffer' fragment. The remaining DNA is termed the 'arms'. Foreign DNA digested with the appropriate restriction enzyme can then be used to replace the stuffer fragment by ligation to the arms. These recombinant fragments can be used for *in vitro* packaging. With insertion vectors (*Figure 3.6*) DNA is inserted at a

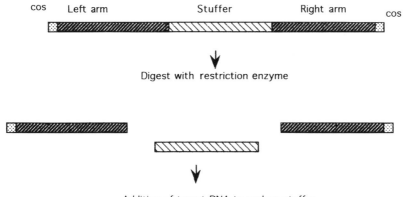

FIGURE 3.5: *Schematic representation of a replacement vector. The arms of the vector and the intervening stuffer fragment are shown. Restriction digestion allows removal of the stuffer fragment. Following ligation of the arms with target DNA at appropriate ratios and concentrations the ligated concatemeric DNA is packaged into phage heads* in vitro.

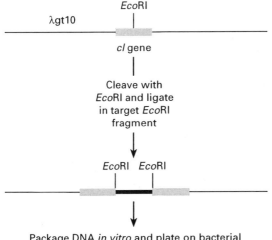

FIGURE 3.6: *Gene cloning in a λ insertion vector. Reproduced from ref. 5.*

single site without removal of vector sequences, thus the size of fragment inserted is usually much smaller than in the case of replacement vectors. One disadvantage of using phage λ vectors for gene cloning has been the difficulty in manipulating the phage DNA, for example isolating arms for use in cloning and the preparation of packaging mixes. However, the advent of the cloning kit has largely elimi-nated the laborious aspects of using bacteriophage vectors. These kits contain all the necessary components for the generation of a library with the exception of the target DNA. Phage λ vectors are also dis-cussed in Chapters 1, 4 and 5.

3.2.3 Cosmids and other bacterial vectors

The development of cosmid vectors is an example of how existing plas-mid and phage λ vectors were imaginatively manipulated to create a vector that shared the most advantageous properties of both [6]. Cosmids are like plasmids in that they have an antibiotic resistance marker and can replicate in *E. coli*. However, unlike plasmids they can accommodate up to 45 kb of DNA. This can be achieved because cosmids contain a cos site (see Section 3.2.2). The presence of this cos site allows vector and target DNA to be packaged as if it were λ DNA. The packaged DNA is then used to infect *E. coli* cells. Once injected the DNA which carries an antibiotic resistance gene can then repli-cate as a plasmid in the cell.

Other vectors include the Gram-positive plasmids which are used for gene cloning in *Bacillus*. They share the same minimal requirements of the *E. coli* plasmids. They have been used extensively for cloning endogenous *Bacillus* genes. Unfortunately, these plasmids tend to be unstable when they have DNA inserted into them [7].

Owing to increased interest in yeast as a host for the expression of for-eign proteins and indeed as a model for studying eukaryotic tran-scription factors, vectors have been developed based on the endogenous 2 μm DNA plasmids or ars (autonomously replicating sys-tems) elements that occur naturally in yeast. Another yeast vector termed the YAC (yeast artificial chromosome) has come to prominence due to its capacity to accommodate very large DNA inserts [8].

3.2.4 Yeast artificial chromosomes (YACs)

Whereas cosmids have a limit of about 50 kb on the size of DNA insert, YACs can accommodate much larger fragments up to 750 kb. They are linear vectors which consist of an origin of replication (ars), telomeric sequences at each end of the vector and a centromere together with a selectable marker gene. These vectors have proved very useful for analyzing genomes by chromosome walking and they have also been used to study gene transcription in large DNA domains.

3.3 Preparation of vector and target DNA

3.3.1 Preparation of vector DNA

Vector DNAs which originate from micro-organisms are propagated in and harvested from their appropriate microbial hosts. However, vector DNAs need to be purified from their hosts in order that they may be manipulated by the variety of enzymes used in gene cloning systems. Most commonly used plasmid vectors are propagated in *E. coli*, though others are grown in other bacteria and yeast.

Typically, a colony consisting of the bacteria which harbor the plasmid vector is used to inoculate a broth of 500 ml. The broth contains an antibiotic for which the resistance gene is encoded by the vector. The bacterial culture is allowed to grow until the substrate has a high optical density. The bacterial cells are then harvested by low-speed (5000 g for 20 min) centrifugation at 4°C. After the supernatant is discarded a large bacterial pellet is left. In order to release the plasmid DNA from this pellet the cells must be ruptured. This can be achieved by a variety of techniques. Sonication and boiling can be used, but by far the most popular method is enzymatic using lysozyme. Lysozyme degrades the peptidoglycan cell wall of bacteria very efficiently. After the cell wall has been disrupted and cell lysis occurs, the plasmid DNA needs to be separated from the cellular proteins and the high molecular weight chromosomal DNA. This can be achieved by treating the lysed cells with the detergent sodium dodecyl sulfate (SDS) (1% w/v) and 0.2 M sodium hydroxide. This SDS/NaOH treament denatures the double-stranded DNA and solubilizes the proteins into the lysate. The alkali treatment is followed by acid treatment with 3 M sodium acetate pH 4.5. After incubation on ice for 1 h, the mixture is centrifuged at high speed (17 000 g) for 1 h. This procedure causes acid precipitation of the protein and high molecular weight chromosomal DNA which are pelleted, leaving the plasmid DNA in the supernatant. The supernatant can then be treated with ethanol and incubated at 0°C to –20°C in order to precipitate the plasmid DNA. The plasmid DNA can then be pelleted by centrifugation (10 000 g for 30 min). After the ethanol has been removed there are a number of ways to proceed for further purification.

One method which can now be considered classical is isopycnic cesium chloride (CsCl) centrifugation. In this case the plasmid DNA is mixed with CsCl solution containing ethidium bromide followed by ultracentrifugation in a fixed-angle rotor for up to 40 h. The plasmid DNA will band at a different density in the gradient than the chromosomal DNA. The covalently closed circular (CCC) supercoiled structure of the plasmid DNA means that it binds less ethidium bromide and so bands denser than linear DNA in a CsCl gradient. The plasmid can be removed from the gradient by puncturing the centrifugation tube, following which the CsCl and the ethidium bromide can be removed by a variety of methods, which include isopropanol extraction and dialysis. More recently protocols involving the use of ion-exchange chromatography available in kit form have circumvented the need for the time-consuming and expensive CsCl ultracentrifugation method and so have become very popular [9].

Isolation of bacteriophage vector DNA is considerably more difficult and time consuming than plasmid DNA isolation. Hence the availability and reliability of 'ready to go' lambda cloning kits which supply the vector 'arms' and packaging mixes have led to a huge reduction in vector preparation by researchers. Several methods are available for isolation of λ vector DNA and these are described in ref. 9.

3.3.2 Preparation of target DNA

In this chapter discussion of the preparation of target DNA will be confined to chromosomal DNA. A description of cDNA preparation for cloning purposes is given in Chapter 4. The isolation of chromosomal DNA to a high degree of purity is necessary for gene cloning purposes. In general, the first stage of preparation is to obtain large quantities of the source material, be it solid tissue, tissue culture cell lines or bacteria. Protocols, of which there are many, will stipulate the amounts of the appropriate material required for a particular protocol. Usually, microgram quantities of purified DNA are required to generate DNA libraries.

The procedures involved in the preparation of chromosomal DNA from mammalian sources and micro-organisms are relatively straightforward which is in contrast to the methods required for preparation of mRNA for cDNA cloning (see Chapter 4). In the case of preparing mammalian DNA, the most basic protocol requires the use of cell rupture techniques using proteinases and detergents to release the DNA, followed by organic extraction of protein by phenol/chloroform treatment, followed by ethanol precipitation of the chromosomal

DNA. When isolating DNA from solid tissue the first step is to flash-freeze the tissue in liquid nitrogen. The tissue is then crushed with a mortar and pestle or a mechanical homogenizer. The macerated tissue is placed in a digestion buffer (100 mM NaCl, 10 mM Tris–HCl, pH 8, 25 mM EDTA, pH 8, 1% SDS, 0.1 mg ml^{-1} proteinase K) where it is subjected to proteinase digestion in the presence of detergent at 50°C for 12–24 h. In the case of tissue culture cells there is no need for this rigorous physical teatment. The cells are treated with the enzyme trypsin to detach them from the plate, followed by centrifugation in order to create a cell pellet, and are then washed. They are then treated with a digestion buffer as described for solid tissue. Following the digestion step the material is treated with the denaturing organic mixture of phenol/chloroform in order to extract the chromosomal DNA. The DNA is retained in the aqueous phase and the proteins are denatured and extracted into the phenol/chloroform organic phase after centrifugation. The DNA is then treated with a salt solution and 2 volumes of ethanol in the presence of which the DNA precipitates and is pelleted by centrifugation at low speed for 5 min. The pellet is washed in 70% ethanol and is then repelleted, followed by resuspension in ultrapure water. If there is any contaminating RNA it can be removed by RNase treatment. The DNA is now ready for modification by the enzymes that were described in Chapter 2.

Extraction of DNA from bacteria follows the same principles as described above. The primary difference is in the lysis procedures required. Indeed, the procedure for rupturing the bacteria is that described in Section 3.3.1 for isolation of plasmid DNA from bacteria.

In this chapter, we have given an overview of the cloning vehicles used in genetic engineering and some methods for isolating these vehicles for experimental procedures. In the following chapters we will describe how they are used in specialized situations, and how they are introduced into their respective hosts and subsequently analyzed.

References

1.   Bolivar, F., Rodriguez, R.L., Betlach, M.C. and Boyer, H.W. (1977) *Gene,* **2,** 75.
2.   Hershfield, V., Boyer, H.W., Yanofsky, C., Lovett, M.A. and Helinski, D.R. (1974) *Proc. Natl Acad. Sci. USA,* **71,** 3455.
3.   Bolivar, F., Rodriguez, R.L., Green, P.J., Betlach, M.C., Heyneker, H.L., Boyer, H.W., Crosa, J.H. and Falkow, S. (1977) *Gene,* **2,** 95.

4. Ptashne, M. (1986) *A Genetic Switch; Gene Control of Phage Lambda*. Blackwell Scientific Publishers, Oxford.
5. Williams, J. Ceccarelli, A. and Spurr, N. (1993) *Genetic Engineering*. BIOS Scientific Publishers, Oxford.
6. Collins, J. and Hohn, B. (1978) *Proc. Natl Acad. Sci. USA,* **75,** 4242.
7. Ehrlich, S.D., Noirot, P., Petit, M.A., Janniere, L., Michel, B. and te Reile, H. (1986) in: *Genetic Engineering* (J.K. Setlow and A. Hollaender, eds), Vol. 8. Plenum Publishing Corp., New York.
8. Burke, D.T., Carle, G.F. and Olson, M.V. (1987) *Science,* **236,** 806.
9. Maniatis, T., Fritsch, E.F. and Sambrook, J. (1989) *Molecular Cloning,* 2nd Edn. Cold Spring Harbor Laboratory Press, Cold Spring Harbor, NY.

4 Generation of cDNA Libraries

Using a term coined by Francis Crick in 1957, the 'central dogma' of molecular biology explains that the hereditary information stored in genomic DNA gives rise to phenotypic manifestation in the form of proteins after interpretation by messenger RNA (mRNA) intermediates. Despite the discovery of the exception to the rule (i.e. the life cycle of retroviruses), we now know that the molecular processes of both eukaryotic and prokaryotic cells reflect this dogma. The process whereby a single-stranded RNA molecule is synthesized from a defined sequence of a double-stranded DNA genome is termed 'transcription'. The enzymes responsible belong to a family called RNA polymerases. The region of genomic DNA used as a template can be called either a 'locus' or, more recently, a 'gene'. Perhaps, as there are RNA molecules that are transcribed from genomic DNA but which are not subsequently translated into protein (e.g. ribosomal and transfer RNA molecules), the term 'transcriptional unit' is most correct.

The term transcription is apt as it defines the process whereby the information stored in genomic DNA is transferred into mRNA intermediates using the same language of nucleic acids. The next step in the genetic flow of information in the cell is 'translation', where the information contained in the mRNA intermediates is used to direct the assembly of proteins from amino acids. Simply, this means that the transfer RNA molecules act as interpreters, taking the instructions stored in the language of nucleic acids and executing them in the language of amino acids, that is the production of proteins.

A definition of a universal transcriptional unit for both prokaryotic and eukaryotic organisms is not possible, although there are similarities between them. The transcriptional unit includes not only the defined region of DNA to be transcribed into RNA but also the regions of DNA recognized by the RNA polymerase before it initiates the transcription reaction. These enzymes recognize very specific regions termed 'promoters' which are located before the region to be transcribed. Analysis of DNA sequences from many promoters shows that there are conserved factors amongst most gene promoters. For prokaryotic promoters, a common region of sequence found 10 bp before the first nucleotide to be transcribed (called position –10) is called the Pribnow box. Another conserved region of DNA sequence is found at –35. This is similar to the basic structure of eukaryotic promoters where a –25 region is called the TATA box and a region at –75 is called the CAAT box. These promoter regions of the genomic DNA are required for recognition and association of the RNA polymerase before it begins to transcribe. The first nucleotide to be transcribed is termed the transcription start site (+1). Once the RNA polymerase binds to a promoter (in reality, this is a complex of several different protein factors), it recognizes one strand of the genomic DNA as a template and it synthesizes a complementary copy using ribonucleotides to form an RNA molecule. This RNA molecule is an exact copy of the nontemplate DNA strand, with the exception of the base uracil (U) being substituted for thymine (T). Convention defines the nontemplate DNA strand as the 'coding strand' and when publishing a gene sequence it is usual to show its nucleotide sequence written 5' to 3' from left to right.

When the process of transcription is studied in more detail the differences between the organization of the transcriptional units in prokaryotes and eukaryotes become more apparent. In prokaryotes, the transcribed RNA molecule is often polycistronic, that is one RNA molecule contains a transcribed copy of the genetic information for several proteins one after the other. This reflects the fact that a typical prokaryotic DNA genome contains relatively little nontranscribed material. This constraint of genome size versus the maximum capacity for genetic information has led to the evolution of operons where genes which define proteins that are associated by their activities are located beside one another and are transcribed from one common promoter. In contrast, most eukaryotic mRNAs are monocistronic. As proteins differ considerably in length, so do mRNA molecules. However, in the late 1970s a strange anomaly was noted. It appeared that the region of genomic DNA coding for an mRNA molecule was far longer than the subsequent transcribed mRNA (e.g. a region greater than 10 kb of chicken genomic DNA coded for the 1.7-kb chicken ovalbumin mRNA). A combination of methods was used to solve this puzzle. Electron microscopy helped provide the dramatic answer. Most eukaryotic genes contain two types of sequence. The first are 'exons', regions of DNA that are found in the mature transcribed mRNA molecule. The second category are 'introns', regions of DNA that are not found in the mature transcribed mRNA molecule. Further analysis showed that both exons and introns are part of the transcriptional unit. In fact, transcription of these 'split genes' produces a long mRNA complementary to both exon and intron sequences. These precursor mRNAs are found in the nucleus and are called heterogeneous nuclear RNA (HnRNA). They are rapidly processed into mature mRNAs by the specific removal of the RNA regions transcribed from the intron DNA sequences. This reaction, called 'splicing', results in a mature mRNA composed only of RNA transcribed from the exon DNA sequences.

A further difference between prokaryotic and eukaryotic mRNAs is the post-transcriptional modification by polyadenylation of most eukaryotic mRNAs. While not fully elucidated, the addition of adenylate residues to the 3' end of transcribed mRNAs (from 20 to more than 100 in length) may have a role in mRNA stabilization. As will be seen, this feature plays an extremely useful role in allowing the study of gene expression using genetic engineering techniques.

If one wishes to study eukaryotic RNA or, by extension, gene expression and protein production, how can one do so? mRNA, unlike DNA, is an extremely labile molecule, sometimes with a half-life measured in minutes *in vivo*. It is also prone to degradation, both chemically in the presence of alkali and biochemically by ubiquitous ribonucleases (RNases). Also, a single-stranded mRNA molecule cannot simply be attached to a vector DNA molecule as when cloning DNA. The obvious answer, considering the ease of cloning DNA molecules for study, is to convert the labile mRNA into more stable double-stranded DNA. This DNA is termed copy DNA or complementary DNA (cDNA). In this form, the cDNA can be manipulated in the same ways as genomic DNA and gene libraries can be constructed with each clone containing a stable copy of an mRNA molecule for further study.

4.1 Preparation of eukaryotic total RNA

The preparation of eukaryotic total RNA is perhaps the most crucial stage in the manipulation of RNA. Preparation of intact RNA, which is intrinsically labile, is further hampered by the ubiquitous nature of RNases. These enzymes are small and maintain their tertiary structure using four disulfide bridges. This gives them the facility to renature, even after treatment with denaturants or after boiling. They also have few co-factor requirements for activity and are active over a wide pH range. Therefore, the molecular biologist must ensure that all reagents and equipment are free of RNase activity before commencing work. In many laboratories, particular chemicals and even bench space are reserved exclusively for RNA isolation. Normal precautions include the use of sterile disposable plasticware rather than glassware, constant use of gloves and some scientists will even use facemasks. Most containers and buffers are also further treated to ensure that they are completely free of RNase. The most common inhibitor is diethyl pyrocarbonate (DEPC), an efficient nonspecific chemical inhibitor. Specific RNase inhibitors are often added to cell lysis buffers to neutralize intracellular RNases released on cell lysis. These include vanadyl ribonucleoside complexes (VRCs). These VRCs act as ribonucleotide analogs and bind to the active site of RNase molecules, inhibiting their degradative activities. In addition, the protein RNAsin is frequently added to inhibit RNases.

The first step in RNA preparation is the collection and handling of cells. Cultured cells may be collected from media by simple centrifugation and, along with tissue samples, are immediately flash-frozen in liquid nitrogen. This is particularly important if cells are to be stored before use. As RNA molecules are labile with short half-lives, it is normal to proceed immediately with RNA isolation. If a storage period is essential then cells are stored at −80°C after being frozen in liquid nitrogen. The cell membranes may then be disrupted using a combination of physical methods, such as boiling, grinding or homogenizing, along with chemical treatments, for example resuspension in denaturing buffers containing detergents, such as sarcosyl or SDS. These denaturant buffers, commonly containing high concentrations (6–8 M)of guanidinium salts or urea, ensure that all the cellular proteins, including RNases, are inactivated [1].

The denatured proteins are then removed by the combination of protein degrading enzymes (Proteinase K or Pronase) and extraction with organic solvents, usually a mixture of phenol and chloroform. These organic solvents precipitate the proteins and allow the efficient purification of the nucleic acids which remain in the aqueous phase. The final step is the isolation of pure RNA. Often, after cell lysis and protein removal, one is left with a preparation of both DNA and RNA molecules. In this case, the DNA can be seen as a contaminating agent with potential to cause severe problems in the generation of cDNA gene libraries, as the genomic DNA can be cloned as easily as mRNA can be converted into double-stranded DNA. Therefore, the normal procedure for RNA purification from a nucleic acid preparation is buoyant density centrifugation, commonly using cesium salts [2]. The large difference in relative densities of RNA and DNA allows their efficient separation and final purification. The purified RNA is finally stored in ethanol at −80°C until use.

4.2 Isolation of poly(A)$^+$ mRNA

Each mammalian cell contains approximately 3×10^{-5} µg (30 pg) of RNA. This population can be subdivided into two main types.

(i)   The RNA products of RNA polymerases I and III (80–90% of total). These RNA products include tRNA and the 18S and 28S ribosomal RNAs which interact with a series of proteins to form the cell's ribosomes, the site of translation.

(ii)   The RNA products of RNA polymerase II (10–20%) include the cell's mRNAs. However, this percentage predominantly reflects the amount of HnRNA present in the nucleus, most of which is degraded in the nucleus giving a final figure of 1–4% mature mRNA found in the cytoplasm.

In quantitative terms, this mature mRNA represents the transcriptional products of approximately 10 000–30 000 different genes at any given time point in any mammalian cell. Some genes are more heavily transcribed, producing more mRNAs, depending on cell type and cell growth phase. One must also consider that only approximately 5% of genomic DNA is transcriptionally active in any one cell. As the mRNA constitutes a minor fraction of the total RNA population, a method to enrich it was devised. As with many genetic engineering techniques, this technique was devised using a natural characteristic of the cell, that is the fact that the vast majority of eukaryotic mRNAs undergo a post-transcriptional modification event with the addition of a series of adenine ribonucleotides to the 3' end. The enzymes involved in this event are the poly(A) polymerases which do not require a complementary strand but simply catalyze addition of adenine nucleotides to the 3' end of template mRNA, triggered by a signal sequence in the mRNA. This structure is routinely called the poly(A) tail.

If a total cellular RNA preparation is passed through a column containing short, single-stranded DNA, constituted of only thymidine residues [poly(T) or oligo(dT)] attached to a solid support (e.g. cellulose) in a high salt buffer, then poly(A)$^+$ mRNA will be retained by the column due to hydrogen bonding between the oligo(dT) residues and the adenylate residues of the poly(A)$^+$ tails (*Figure 4.1*) [3]. The vast majority of the nonpolyadenylated cellular RNA passes directly through the column. The poly(A)$^+$ mRNA can then be eluted directly from the column by the addition of low salt buffers which destabilize the hydrogen bonding. Typically, 1–2% of the total RNA added to the

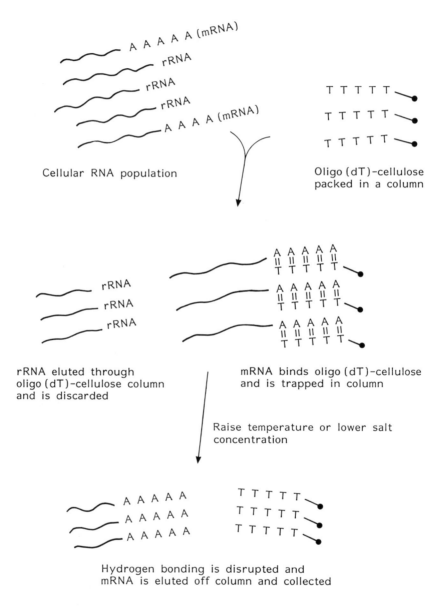

FIGURE 4.1:   *Purification of messenger RNA by oligo(dT) chromatography.*

column can be recovered as poly(A)$^+$ mRNA. This is precious material and is stored in ethanol at $-80°C$ prior to use.

4.3 Conversion of mRNA to cDNA

At the beginning of this chapter, it was mentioned that there were natural exceptions to the central dogma of molecular biology elucidated by Francis Crick. One of these exceptions is the life cycle of the retroviruses. These viruses have a genome of RNA rather than DNA. Upon infection of eukaryotic cells, they have the ability to produce a double-stranded DNA copy of their RNA genome (see *Figure 4.2*). This DNA copy has the ability to integrate into the host cell genome and remain dormant. The enzymes involved in the synthesis of DNA from an RNA template are the reverse transcriptases (RTase). In a manner which emerges as the theme of this book, the genetic engineers have made their own use of the abilities of this enzyme to convert *in vitro* labile mRNA into more stable cDNA for further analyses.

RTase, in a similar fashion to DNA polymerases, can be used to synthesize a DNA molecule complementary to an mRNA template. The enzyme requirements are the presence of deoxynucleotide triphosphates, its optimal activity buffer, and a double-stranded primed region on the RNA template. This double-stranded primer region is easily prepared. Short single-stranded oligo(dT) fragments are added to the poly(A)$^+$ mRNA preparation. The oligo(dT) fragments hydrogen bond with the poly(A) tail to form the double-stranded primer for the RTase. After recognition and binding of the RTase to this site, the polymerase proceeds to synthesize a DNA strand complementary to the RNA molecule obeying the 5'–3' directionality of nucleic acid synthesis.

The result is a double-stranded 'hybrid' molecule, one RNA strand and one DNA strand. There are reports that such a molecule is stable enough for joining to vector molecules and introduction into host cells but the number of recombinant clones generated in this fashion is usually quite low. Conversion of this hybrid molecule to a double-stranded DNA molecule by replacement of the mRNA template strand was initially performed by chemical degradation [4] but nowadays is normally performed enzymatically. Chemical degradation of the RNA is carried out by the addition of alkali (0.1 M NaOH) to the reaction. The ribonucleotides are cleaved but the deoxyribonucleotides remain

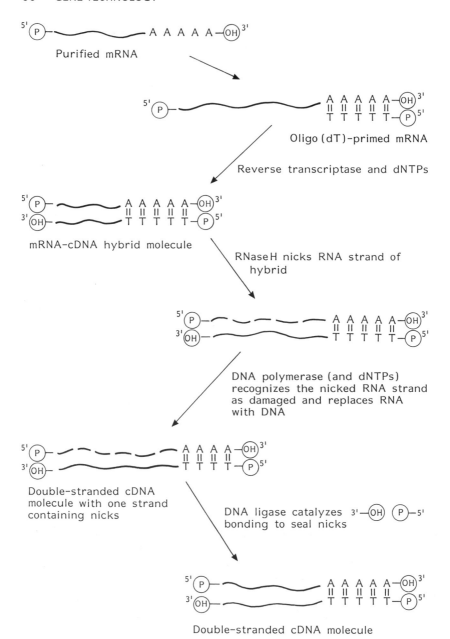

FIGURE 4.2: *Enzymatic construction of double-stranded cDNA molecules from mRNA.*

intact. This results in a single-stranded DNA molecule. Because of the propensity for even one or two complementary bases to hydrogen bond with one another, when the alkali is neutralized (addition of 0.1 M

HCl) the 3' end of this DNA molecule forms short transient structures called 'hairpin loops' which provide temporary double-stranded primers for the action of DNA polymerase I to synthesize a second strand of DNA complementary to the first DNA strand. The result is close to the desired stable double-stranded DNA copy of the original mRNA template. However, it remains in a form that cannot be cloned as one of its termini is a loop. In this form the fragment cannot be joined to a vector molecule in a ligation reaction. The loop, which in reality is a small single-stranded region, must be removed. This can be performed enzymatically by the addition of the enzyme nuclease S1 which has a preference to degrade single-stranded as opposed to double-stranded DNA. The resulting double-stranded DNA molecule is now ready for ligation to a vector molecule before introduction into host cells to generate a cDNA gene library. However, a drawback to this method is that as the nuclease S1 degrades the single-stranded hairpin loop, it is actually removing some of the DNA originally transcribed from the mRNA by the RTase. Therefore, this method produces cDNAs that are not exact copies of the mRNA, that is they may not be full length. The missing DNA sequences might prove vital to further analysis or research.

Faced with this disadvantage, a second method using enzymatic degradation of the mRNA strand with no degradation of the DNA strand was devised and is commonly called the 'RNaseH' method [5] (see *Figure 4.2*).

Initially, the oligo(dT) priming and RTase action is identical to the older method. The removal of the mRNA strand is by the action of the enzyme RNaseH. This endoribonuclease cleaves the phosphodiester bonds joining adjacent ribonucleotides when RNA is in a double-stranded structure with DNA as is formed from the RTase step. These 'nicks' in the RNA strand are, advantageously, recognized by DNA polymerase I as damaged nucleic acid and this enzyme will bind to these regions of the hybrid molecule. On doing so, it also recognizes the hybrid molecule as artificial (as it is!) or damaged DNA, and begins to remove the RNA using its associated exonuclease activity, meanwhile substituting deoxyribonucleotides and synthesizing DNA using the other DNA strand as template. In performing this action, the DNA polymerase is only carrying out one of its natural roles in the cell for DNA maintenance and removal of RNA primers that aid genome replication. The result is a molecule similar to that after RNaseH treatment except that it now is composed of two DNA strands. The nicks remain, however, as the DNA polymerase cannot catalyze the formation of the phosphodiester bond between two adjoining deoxyribonucleotides (this would be a 3'–5' bonding). However, these nicks can be sealed by the action of DNA ligases. The final result is a double-stranded DNA copy of the original mRNA with no sequences removed. One more manipulation is required before this molecule can be ligated to plasmid or phage vectors and introduced into host cells generating a cDNA library.

As discussed in Chapter 2, the joining of two DNA fragments occurs far more efficiently as a cohesive ligation. This requires that both DNA fragments (e.g. cDNA and vector molecules) have single-stranded DNA protrusions that are complementary. They will then have the ability to associate by hydrogen bonding before being sealed together as a new recombinant DNA molecule by a DNA ligase. The cohesive ends at the termini of the vector are determined by the choice of restriction enzyme used to prepare it for the cloning experiment. Logically, if one could add similar complementary termini to the cDNA molecules, an efficient cohesive ligation could be peformed. Such modification of the termini of cDNAs can be performed by the addition of short synthetic double-stranded DNA fragments to the ends by blunt-end ligation using a DNA ligase (see *Figure 4.3*). These synthetic DNA fragments are called linkers or adaptors. Usually 12–20 bp in length, they are designed so they carry internally an appropriate restriction enzyme recognition site. Therefore, appropriate linkers are ligated to the cDNAs. This blunt-end ligation can be performed efficiently as a vast numerical excess of linker fragments can be added ensuring that both termini of each cDNA receive a linker molecule (in practice, a series of linker molecules are ligated to each end). The cDNA–linker molecules are then digested with the appropriate restriction enzyme resulting in cDNA fragments with new cohesive termini. Now, an efficient cohesive ligation reaction can be performed between vector and cDNA molecules ensuring that an optimal number of new recombinant cDNA molecules are ready for introduction into host cells.

Proving that nothing is as simple as it sounds, there is another enzymatic manipulation of the cDNA–linker molecules that should be considered before the ligation to the vector molecules. While the choice of restriction enzyme and appropriate linker is made to suit the cDNA cloning strategy, one variable usually remains. It is highly likely that the sequence of the mRNA(s) of interest is unknown — the determination of the sequence may be the final research goal. Therefore, the researcher must choose suitable restriction enzymes with no knowl-

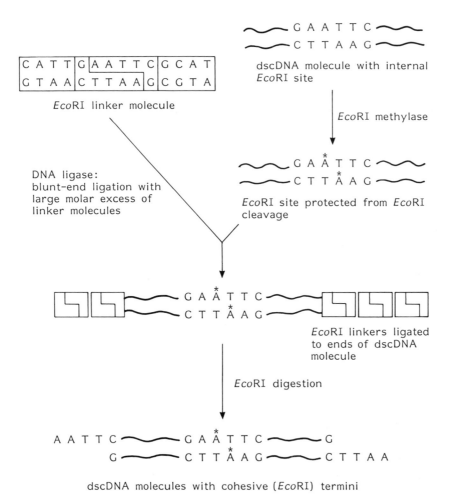

FIGURE 4.3: *Addition of cohesive termini to double-stranded cDNA molecules using linkers.*

edge of whether the cDNA molecules generated will contain recognition sites for these enzymes. If the mRNA of interest does contain these sites, then so will the cDNA and, upon digestion of the cDNA–linkers, the cDNA will also be digested into two or more fragments. While these fragments may be ligated individually to vector molecules, the identification of all these individual clones and the stitching together of their sequences to form a complete study is a very onerous if not impossible task. To resolve this problem, all recognition sites found internally in the cDNA must be protected from the action of the restriction enzyme. This is performed by treating the cDNA, before ligation to linker molecules, with methylase enzymes (see Section 2.2.4). These enzymes protect the bacterial genome from its own restriction enzymes by methylating specific bases in the recognition sequence rendering it resistant to digestion. Therefore, internal sites in the cDNA can be protected and will not be digested by the action of the restriction enzyme creating single-stranded termini from the linkers.

4.4 Strategies for cDNA libraries

Taking into account the preceding sections, the isolation of quality poly(A)$^+$ mRNA should allow the generation of full-length cDNA molecules. This is particularly so for mRNA molecules between 1 and 5 kb in length. It is sometimes more difficult to generate full-length cDNA from larger mRNAs. Experimentation and optimization of the RTase step may be required, together with a little luck. There are two main issues to consider when one must decide on a strategy to generate a cDNA library. The first issue is the relative abundance of the mRNA(s) of interest and the second issue is the subsequent screening method to be used to identify clones of interest from the library.

4.4.1 mRNA abundance

The number of mRNA molecules per cell transcribed from a particular gene is termed its abundance. Analysis has shown that, in general, the mRNA population of a eukaryotic cell can be separated into two or three classes based on relative abundance and ranging from highly abundant to rare. The highly abundant mRNAs may constitute up to 50% of the total mRNA. However, this class may be made up only of the transcriptional mRNA products of two or three genes (e.g. ovalbumin, conalbumin and lysozyme from oviduct cells or hemoglobin in erythroid cells). The middle abundant class may correspond to the mRNA products of 8–10 genes with expression levels of approximately 1%, while the rare class can be composed of the mRNA products of 10–20 000 genes making up approximately 30% of the total mRNA. As a rule of thumb, the highly abundant and abundant mRNAs tend to show tissue specificity, that is their genes are only expressed at high levels in certain tissues and not at all or at basal levels in other tissues. They are thus related to specialized cell functions. In contrast, there is often overlap between the rare mRNAs from many different tissues. The term 'housekeeping' genes has been coined for these genes, meaning that their transcription is required by all cells, for example enzymes for the maintenance of the cell's metabolism. These differences amongst a cell's mRNA population reflect the different strengths of gene promoters and also different stabilities of individual mRNAs. They give the genetic engineer some flexibility in designing strategies for the generation of cDNA libraries. If an mRNA of interest is abundant, then the efficiency and number of clones to be generated for the library is not critical (e.g. with an mRNA of 10% abundance, a gene library of 100 clones should contain 10 coding for that mRNA). This allows a greater choice of vector molecule to introduce the cDNA into the cell. Plasmids, for instance, with their relatively poor efficiency for transfer into cells when compared with phage vectors, might well be the vector of choice for cloning abundant mRNAs. Perhaps, more routinely, the abundance of an mRNA of interest is not known or is thought to be rare. In these cases, it is wise to design a cloning strategy that will optimize the chance of producing cDNA gene libraries with several times more clones than the number expected. Typically, these libraries should contain 10^5–10^6 individual clones, or expressed in another manner, 10^2–10^3 times the average number of genes expressed. One can feel confident that the screening of such numbers of cDNA clones will result in the identification of the clone of interest.

4.4.2 Library screening

Of course, once a cDNA gene library has been constructed, the problem remains to identify particular clones of interest from the library for further study, including the elucidation of their DNA sequence. While there are several methods available to 'screen' gene libraries, they can be characterized into two groups. The first is the use of nucleic acid probes and the second approach is the use of antibody probes. Any strategy to clone cDNA must take into account the subsequent screening method, as specialized vectors may be required for a particular method. Nucleic acid probes, DNA fragments, RNA molecules and synthetic oligodeoxynucleotides, do not require specialized vectors and are often the screening method of choice. Antibody probes do, however, require specialized vectors. In this case, one is dependent on the cDNA fragment being transcribed and translated in the host cells so that the recombinant protein can be identified by the antibody. Generally, these specialized vectors, termed expression vectors, allow the cloning of cDNA fragments at an internal site in a vector gene coding for a highly expressed host protein (e.g. β-galactosidase in *E. coli*).

Therefore, a fusion protein from both the host gene and cloned cDNA may be produced and act as a target for the antibody probe. One should note that cDNA libraries to be screened in this manner should theoretically be six times larger in numbers of clones to cover the 1 in 2 chance of a particular cDNA clone being in the correct orientation with respect to the host gene promoter and the 1 in 3 chance of the cDNA clone being in the correct triplet reading frame for translation with the host gene protein.

Taking all of the above into account, cDNA library generation and screening have now become a routine aspect of molecular biology research. They are the start point of most projects to analyze gene function, either by the use of the characterized cDNA as a probe which is used to isolate the genomic DNA, or its use as a probe to monitor gene expresssion in different physiological situations. Of course, the fact that the cDNA may be expressed as a protein has tremendous practical consequences and has been one of the cornerstones upon which modern biotechnology has been built. Indeed, it has been said that the old dogma of DNA = RNA = Protein has been replaced by a new dogma of DNA = Protein = Dollars!

References

1. Glisin, V.R., Crkvenjakov, R. and Byus, C. (1974) *Biochemistry*, **13**, 2633.
2. Maniatis, T., Fritsch, E.F. and Sambrook, J. (1989) *Molecular Cloning*, 2nd Edn. Cold Spring Harbor Laboratory Press, Cold Spring Harbor, NY.
3. Aviv, H. and Leder, P. (1972) *Proc. Natl Acad. Sci. USA*, **69**, 1408.
4. Higuchi, R., Paddock, G.V., Wall, R. and Salser, W. (1976) *Proc. Natl Acad. Sci. USA*, **73**, 3146.
5. Gubler, U. and Hoffman, B.J. (1983) *Gene*, **25**, 263.

5 Introduction of Recombinant DNA Molecules into Host Cells

If the ability to manipulate nucleic acids and engineer the construction of new 'recombinant' DNA molecules is the first prerequisite for the discipline of genetic engineering, then the second and no less important prerequisite is the development of techniques to introduce recombinant DNA molecules into living cells. The result of the introduction of recombinant DNA molecules into cells is the generation of 'clones', that is individual cells containing one recombinant DNA molecule that can be both propagated and stored to produce suitably large quantities of the recombinant DNA molecule for study and to conserve the recombinant DNA molecule in a stable form. There are two salient features of these 'cloning' experiments. First, each DNA fragment will be introduced into, and stably inherited in, a single host cell. This allows the separation of individual DNA molecules from a population mixture such as double-stranded cDNA produced from mRNA and genomic DNA fragments produced from restriction enzyme digestion. Therefore, the introduction of these heterogeneous populations of DNA fragments into host cells results in the generation of 'gene libraries', where each host cell will stably maintain one DNA fragment from the original mixture. This separation of DNA molecules from the mixture is critical to the ability to identify and isolate a particular DNA molecule (e.g. a gene) from the genome of interest. Secondly, an advantageous feature is the possibility of the growth of the host cells, by simple culture, to produce suitably large quantities of the recombinant DNA molecule for study. For instance, one *E. coli* cell containing one recombinant DNA molecule can be cultured overnight to form a colony on an agar plate. This colony represents over 10 million cells each containing at least one molecule of the recombinant DNA. The inoculation of a 500-ml liquid broth with this colony will produce, after 24 h, the equivalent of 5 x 10^{10} cells containing at least 5 x 10^{10} recombinant DNA molecules. Therefore, in 1–2 days, one can quite easily produce milligram quantities of recombinant DNA sufficient for further analysis such as DNA sequencing.

In this chapter two aspects of genetic engineering are described. The first is the description of techniques to enable the recombinant DNA to traverse the cells' membranes and/or walls. For eukaryotic host cells which are used in special situations, this process is further complicated by the need for the recombinant DNA to pass through the nuclear membrane to reach the site of the genome where replication occurs. The second aspect is the fate of the introduced recombinant DNA (more commonly referred to as foreign DNA). The aim is to have it stably inherited, becoming part of the host's genome and so passed on to all progeny cells. Both these factors create separate problems, and the genetic engineer must design an experimental system to overcome both. While such systems were developed many years ago for the 'work horse' of genetic engineering, *E. coli*, it is now possible to introduce foreign DNA molecules into plant and animal cells with relatively good prospects of a stable inheritance. With the various hosts, different methods for introduction of foreign DNA are required. Some, particularly for bacteria, are simple in design whereas others require specialized equipment (e.g. microscopes and micromanipulators for the generation of transgenic animals). All have different efficiencies of success, that is the number of new clones produced. The general term for the process is 'transformation', originally used for bacteria but now used to describe the introduction of DNA into any bacterium, fungus, animal or plant cell.

5.1 Introduction of foreign DNA into bacteria

Many bacteria have a natural ability to take up free or naked DNA into their cell. The DNA can be a linear fragment or a circular molecule and can be of varied length, possibly containing many genes. In nature, this is a method of evolution for micro-organisms, conferring on them the ability to generate greater genetic diversity by incorporating new genes and therefore produce new proteins and phenotype. Originally thought not to be a major source of new genetic information used by bacteria in nature, the increasing description of different bacteria utilizing this mechanism has led to more study of the importance of this natural ability.

Two of the most fundamental experiments in biology or genetics, where DNA was proven to be the cell's molecule of genetic information, used the natural ability of bacteria to be transformed with free DNA. In the 1920s, a British health officer named Griffith [1] discovered the process of transformation. Studying the virulence of different strains of the bacterium *Streptococcus pneumoniae* in mice, he showed that virulent *S. pneumoniae* strains that he had experimentally killed by heat treatment, could, after injection, kill mice and be recovered from the dead animal. This result only occurred if he injected the mice with a mixture of the heat-killed virulent cells and cells from an avirulent strain of *S. pneumoniae*. Only virulent cells were recovered from the dead mouse. What had happened was that somehow the avirulent cells had become virulent, that is become 'transformed'. In the 1940s, Avery, McCarthy and McLeod showed that this transformation by *S. pneumoniae* cells was due to the uptake of DNA, and not some other cell molecule such as protein or carbohydrate, by the avirulent cells from the heat-killed virulent cells. These were the first demonstrations that DNA is the molecule of genetic information and heredity in the cell.

The ability of a bacterium to be transformed is dependent on its competence, that is its ability to take up DNA from its environment. With some bacteria, competence is dependent on the stage of growth of the cell, for example *S. pneumoniae* cells develop competence late in logarithmic growth. Of more interest to the genetic engineer is the ability to manipulate this competence experimentally, and techniques now exist to make many bacteria competent by growing them in defined media or washing them in defined chemical solutions.

As indicated earlier, the problem of stable inheritance of foreign DNA in bacteria has also been overcome. Simply by joining foreign DNA to vector DNA molecules such as plasmids and bacteriophage, which have an inherent ability to replicate in bacteria, one can assure that foreign DNA once introduced will survive and persist in all progeny cells. Therefore, methods for the introduction of foreign DNA into bacteria are designed primarily for the introduction of plasmids and phage into the cell and often mimic the processes by which these occur naturally.

5.1.1 Transformation of *E. coli* with plasmid molecules

As with many other breakthroughs in genetic engineering, the development of transformation systems for *E. coli* happened in the early

1970s. Although thought to be refractory to transformation, in 1970 Mandel and Higa showed that *E. coli* cells treated with calcium chloride had the ability to take up naked bacteriophage lambda DNA [2]. Subsequently, in 1972, Cohen and his co-workers showed that $CaCl_2$-treated cells were also competent to receive plasmid DNA [3]. Now, almost any strain of *E. coli* can be transformed using this treatment. Efficiencies of transformation (i.e. the number of transformed colonies produced using a known amount of plasmid DNA) can vary. One of the reasons for differing efficiencies is that bacteria have natural defense systems to protect themselves from invading DNA (e.g. phage DNA). These are called restriction systems and involve the production of restriction enzymes (see Chapter 2). This natural defensive ability of wild-type bacteria against foreign DNA is a disadvantage to the genetic engineer whose recombinant DNA would be destroyed once introduced into the cell. The obvious solution of generating mutant strains, deficient in their restriction systems, was used to select *E. coli* hosts for genetic engineering cloning experiments. Also, these hosts often have mutations in their recombination systems to ensure that the foreign DNA introduced retains its integrity and does not undergo any recombination events with the host genome or become lost to the cell.

The mechanism by which *E. coli* cells become competent to plasmid DNA may not be fully clear but has been well examined and is quite a simple laboratory protocol (*Figure 5.1*). It has been shown that *E. coli* cells and plasmid DNA interact in the presence of calcium ions and at low temperature (0°–4°C). These Ca^{2+} ions may promote cell membrane changes that facilitate DNA transfer into the cell. A subsequent heat shock (37°–42°C for 2–5 min) appears to promote the transfer of the plasmid molecule into the organism. Nutrient medium is then added to allow the cell to recover from the salt solution and also to allow the plasmid to replicate and transcribe and translate its genes into protein. One of these gene products of the plasmid will be a selectable marker allowing cells that have taken up the plasmid to be distinguished from those that have not. This selection is critical as analysis has shown that only approximately 1% of *E. coli* cells are actually transformed with plasmid DNA. The reason for this low percentage remains unknown. Typically, at least one million clones can be obtained when $CaCl_2$-treated cells are transformed with 1 µg of plasmid DNA. If this plasmid DNA has been manipulated and is a recombinant molecule, that is another molecule has been ligated to it, the transformation efficiency may fall to 10^5 clones per microgram of plasmid DNA. In 1983, Hanahan showed that the use of other cations

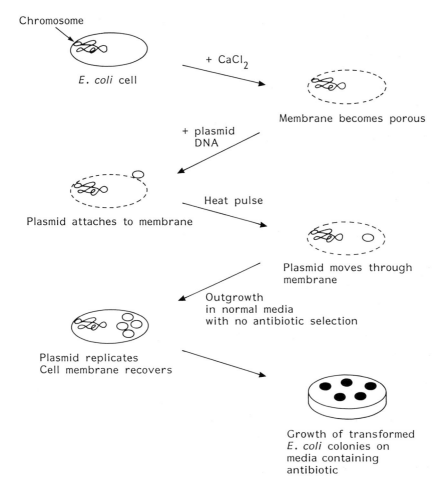

FIGURE 5.1:   *Transformation of* E. coli *cells with plasmid DNA.*

particularly rubidium chloride, along with optimization of other steps in the protocol, could produce efficiencies of 10^8 clones per μg of plasmid DNA [4]. These efficiencies are sufficiently high to promote confidence that every DNA fragment in a population mixture for cloning will end up stably inherited in a bacterial cell. Such a probability is critical as interest is usually only in one fragment (e.g. a specific gene) of the population mixture. The high efficiencies using divalent cation-treated *E. coli* cells and plasmid molecules as vectors in theory allow the choice of plasmid as a vector for cloning even the most complex populations (e.g. mammalian genomic DNA libraries and cDNA libraries). However, plasmids are not a routine choice for cloning such complex populations. The reason is primarily a practical one. The storage of 10–100 million individual bacterial clones, each containing a recombinant DNA fragment, to maximize the probability that one has cloned a complete mammalian genome or a complete mRNA population, is extremely laborious. Therefore, plasmids are primarily used to clone smaller genomes (bacterial or fungal) where 10 000 clones, each containing up to 10 kb of foreign DNA, easily cover the genomes of study.

5.1.2 Transformation of *E. coli* with bacteriophages

Bacteriophages are more common vectors for the generation of cDNA and eukaryotic genomic DNA libraries than are plasmids. This is due to their ability to accept larger DNA molecules into the cell at higher efficiencies than plasmids. Unlike plasmid vectors whose cloning capacity is about 10 kb, certain phage vectors (i.e. replacement vectors) can accept 20–30 kb of foreign DNA while still being able to infect cells at high efficiencies [5]. Bacteriophage lambda is the most common phage vector of *E. coli*. Several modified lambda phages exist for cloning purposes. As for plasmid transformation, the host *E. coli* strains have also been manipulated. *E. coli* mutants which have a high propensity for infection by lambda are used. A recombinant phage DNA molecule can of course be introduced into *E. coli* cells just as plasmids, by treatment with $CaCl_2$. However, the efficiency is approximately 10-fold less than that of plasmids making it inefficient. The optimal method for introduction of phage DNA is to mimic the natural phage infection mechanism for *E. coli*, that is to use the whole phage rather than phage DNA to infect cells. Therefore, one must create recombinant phage from recombinant phage DNA. This is possible due to our understanding of the actions and life cycle of lambda phage, long studied by microbiologists and bacterial geneticists before the advent of genetic engineering. The complete nucleotide sequence of lambda phage (length approximately 50 kb) is known as are the proteins that are coded and their functions [6]. The temporal sequence of expression of these proteins after infection of *E. coli* cells is also known. Most of this work involved the creation of lambda mutants deficient in the production of various proteins. This knowledge, along with some of the mutants, has given the genetic engineer a relatively simple method for the production of lambda proteins. These proteins if added to parental or recombinant lambda DNA will react with the DNA to form viable lambda phage. This process where the lambda DNA is packaged inside a protein coat to form phage is termed '*in vitro* packaging' [7].

How does one produce these lambda proteins required for packaging? Two particular *E. coli* strains are frequently used [8]. Both share a common characteristic and that is they are both lambda lysogens. As explained earlier, lysogen is a term for an *E. coli* with a lambda genome integrated into its chromosome. This lambda is termed a prophage and the lambda is said to be in its lysogenic life mode rather than its lytic mode (lysogeny is analogous to viral latency stages). In this mode the lambda phage remains dormant until activated by a signal when it removes itself from the chromosome and re-enters the lytic mode. These particular *E. coli* lysogens also have a specific muta-tion in their lambda prophages, whereby the signal to change the prophage from lysogenic to lytic modes is a simple heat shock (e.g. 15 min at 45°C rather than the normal 37°C). Once the lambda has removed itself from the *E. coli* chromosome, it quickly replicates itself to 1–200 copies inside the cell and then begins to transcribe and translate the necessary lambda proteins to form viable phage. Normally, these phage would then lyse the cell and proceed to infect others. However, in the case of both these prophages there is another mutation that blocks the formation of one lambda protein necessary for packaging (in both prophage a different protein-coding gene is mutated). As not all the lambda proteins are available, no viable phage are formed after heat induction of the lytic mode in both these *E. coli* lysogens. Instead, the cell becomes full of lambda proteins. Therefore, the genetic engineer can prepare large amounts of these proteins by growing up 500-ml batches of both lysogens at 37°C, fol-lowed by heat induction and further growth for 4–5 h to allow trans-lation of the lambda proteins. The cells are then harvested and lysed and the lambda proteins separated by centrifugation from the genome and other cellular material. If the lambda protein preparations from both lysogens are then mixed, they will complement one another and all the required lambda proteins will be available for *in vitro* packag-ing of recombinant lambda DNA. The process is as simple as mixing the two components in the correct quantities, or, nowadays, purchas-ing the packaging mixes. All the prepared lambda protein extracts required in order to package lambda molecules to form recombinant phage is the presence of the 'cos' sites which are naturally part of lambda. The cos sites should be separated by a distance of 45–52 kb (roughly the size of the wild-type lambda genome).

These recombinant phage are then used to infect *E. coli* cells. The cells are grown in the presence of magnesium ions to support phage particle integrity and maltose, which induces the production of mal-tose receptors on the cell membrane. The cells are then mixed with the phage for 30 min before plating out on nutrient plates. The site of attachment of lambda phage to the *E. coli* cell is the maltose receptor and so optimal efficiencies of infection are obtained. These efficiencies are 10^8–10^9 clones per µg of parental lambda and approximately 10^7 clones for recombinant lambda phage. The clones are seen as clear plaques on lawns of *E. coli* growth. Overall, the efficiency of clone production is 100-fold greater than $CaCl_2$-mediated transformation with plasmids. As lambda vectors can incorporate at least twice the size of fragments of DNA as plasmids, there is a potentially 200-fold greater cloning capacity using lambda vectors. Also, gene libraries containing 10^6–10^7 phage do not require individual isolation of each clone for storage. The phage library can be stored as a mixture for years.

5.1.3 Transformation of *E. coli* by electroporation

The use of high-voltage electric pulses to introduce plasmid DNA into *E. coli* and other bacterial species along with eukaryotic cells including mammalian eggs is rapidly gaining acceptance as a useful technique. A specialized energy source and cuvettes containing built-in electrodes are required, but the high transformation efficiencies (e.g. 10^{14} clones per µg of plasmid DNA with *E. coli*) produced by electroporation have led to its use in many molecular biology laboratories. The procedure is as simple as $CaCl_2$-mediated transformation. Cells are grown to a certain stage (this may vary with different species), washed with sterile water to remove nutrient, mixed with plasmid DNA and submitted to an electric pulse of about 2500 V for 3–5 msec. While many bacterial cells are killed by this treatment, many survive and take up plasmid DNA [9]. Electroporation has proved extremely useful in developing methods for DNA introduction into other bacterial species and eukaryotic cells, in particular plant cells.

5.1.4 Transformation of other bacterial species

The study of transformation systems for other bacteria, while lagging behind *E. coli,* is expanding. This is particularly true for bacteria which produce useful products such as antibiotics, amino acids and enzymes. The production of larger quantities or modified products by these bacteria might be facilitated by the ability to engineer these strains genetically. This has led to increased interest in genetic manipulation of other bacteria. In many cases, the search for new transformation systems is directly analogous to those of *E. coli,* that is to find suitable plasmids and phage for a bacterium of interest and determine methods for their introduction. The use of divalent cations such as Ca^{2+} is documented as useful for many other bacteria, particularly Gram-negative bacteria. The problem with Gram-positive bacteria is the fact that they have a more complicated cell wall structure which is not easily amenable to DNA transfer into the cytoplasm. However, research on *Bacillus* species, from which many enzymes are produced commercially, has shown that some are naturally competent. In addition, enzymatic stripping of the cell wall of industrial isolates to form protoplasts can allow efficient uptake of DNA. Eventually, after DNA uptake, the complete cell wall can be regenerated. Now, electroporation has been shown to be a most useful technique in transferring DNA into different bacteria [10]. These include *Clostridium, Bordatella* and *Corynebacterium* species. Often, all that is required is the optimization of the electric pulse for a given bacterium, such that maximum numbers of cells are transformed and minimum numbers of cells are killed.

5.2 Introduction of foreign DNA into eukaryotes

One might first ask: why introduce foreign DNA into eukaryotic cells or organisms when the use of bacterial hosts such as *E. coli* is apparently uncomplicated and successful? The main reason is for the production of recombinant eukaryotic proteins. It quickly became apparent that although eukaryotic genes could be cloned into bacteria and manipulated to produce recombinant protein, in many cases the protein did not exhibit its correct biological activity, that is its activity was modified compared with the native protein. It was found that recombinant eukaryotic proteins produced in bacteria are often incorrectly folded and also do not undergo post-translational modifications such as glycosylation which would occur normally. Also, many functional recombinant proteins are not secreted by bacteria which lack appropriate secretion mechanisms. Not only is this a disadvantage in the purification of recombinant protein but often these nonsecreted recombinant proteins collect in exclusion bodies (proteinaceous particles) in the bacterial cell. These protein aggregates can precipitate killing of the host cells and also become insoluble rendering themselves useless for further application. The inability of bacteria to splice the introns of eukaryotic genes was also a reason for investigating the use of eukaryotic hosts for the production of correctly functional recombinant eukaryotic proteins. The cause of the protein problem was thought to be solely the host's prokaryotic background.

Therefore, a possible solution would be to clone into a eukaryotic environment with the obvious candidate being the least-complex but well-studied eukaryote, yeast. It is fair to say that this solution has presented as many problems as the original. However, the result has been to further research using more complex hosts such as mammalian cell lines. As a result, today it is possible to produce recombinant eukaryotic proteins in a host almost identical to the natural host, that is transgenic plants and animals.

5.2.1 Transformation of yeast

The original candidate eukaryotic host was the yeast, *Saccharomyces cerevisiae*. Used commercially in the production of bread and beer, it is one of the best-studied eukaryotes; the '*E. coli* of the eukaryotic kingdom'. Its genome is approximately 2×10^7 bp in length, only four-fold larger than *E. coli* and 150-fold smaller than the human genome. The splicing of mRNA is generally not species-specific and so yeast cells have the ability to splice intron-containing transcripts from many types of genes and can also secrete proteins into the medium for easier collection and purification. Unlike bacteria, the genome of *S. cerevisiae* is divided into linear chromosomes. However, it does contain a type of plasmid called the 2 μm circle. This molecule has its own origin of replication (ars) and so can replicate independently of the chromosomes.

The method used for the introduction of foreign DNA into yeast cells is similar to that used for some Gram-positive bacteria. The cell wall must be enzymatically degraded to form protoplasts (which must be kept in isotonic medium to avoid cell rupture) and the protoplasts are then treated with calcium ions to allow the uptake of foreign DNA. Intact cell walls can then be regenerated by growing the treated protoplasts on specially supplemented media.

Yeast vectors have also been developed to maintain foreign DNA in all yeast progeny. These plasmid-like molecules usually contain a yeast origin of replication either from the 2 μm plasmid or from autonomously replicating systems (ars) found in all yeasts. The vector will also contain a selectable marker (analagous to antibiotic resistance genes on bacterial plasmid vectors) to distinguish clones from yeast cells without the vector. In most cases this is a metabolic gene producing an enzyme for which the host cell is deficient. This disabled host (termed an auxotrophic mutant) cannot grow unless its mutated gene is complemented by the functional gene present on the vector. Therefore, only host cells containing the plasmid can grow.

In an exciting development in molecular biology and genetic engineering over the last few years, yeast vectors called yeast artificial chromosomes (YACs) have been developed. As described previously in Chapter 3, these vectors contain yeast centromeric and telomeric sequences (the middle and end portions, respectively, of yeast chromosomes). Large fragments of DNA up to several hundred kilobases in size can be inserted between the centromere and telomere sequences to form an artificial yeast chromosome. This can then be introduced into yeast cells. The potential to separate such large DNA fragments and the ability to clone a complex genome (such as the human genome) with a library of only several thousand independent clones using YACs has greatly facilitated research into mapping and ultimately sequencing the human genome.

5.2.2 Animal cells

Although all routine gene isolation (cloning) is carried out in simple host systems (bacteria and yeast), some projects require the use of complex animal cells grown in culture. Many, but not all, animal cells can be grown *in vitro* and the choice of cell type is dependent on the reason for their use. In Chapter 7 we will see that analysis of the isolated gene frequently passes through a phase that involves transfer to selected animal cells. For the purposes of the present chapter, it is appropriate to indicate that animal cells grown in culture are frequently needed when large amounts of protein are required, for example in the production of therapeutic products by the biotechnology industry. The reason why this more costly route is taken is because bacteria lack some of the enzymes that are needed for the correct modification of the protein (e.g. glycosylation). The properties of the protein and its interaction with the patient can be altered dramatically by such subtle changes. Therefore, the need for biologically active products dictates the choice of animal cells for protein production in these circumstances.

The DNA constructs that are used for this purpose are different from those used for bacteria which reflects the special requirements of animal cells. The origin of replication, the promoters and the selective markers of animal cells are different from those from bacteria. Most vectors used to date include components that have been well studied in DNA viruses. These provide not only an origin of replication but also strong promoter regions which allow the expression of the gene of interest to take place. Several methods exist to achieve the transfer of the vector into the animal cell [11]. These include the use of divalent cations (calcium phosphate), electroporation and liposome-mediated delivery; the development of viral vectors is also a very active area of research. Following the transfer of the vector it will reside as a multi-copy episome in the cytoplasm of the animal cell for several days (generations). During this time it is possible to test whether the sequence is expressed. When the cell lines that express the sequence are obtained it is then useful to allow stable transformants to emerge. This follows the integration of the DNA in a random manner into the animal cell genome. The consequences of this integration are dependent on the locus into which the DNA integrates; frequently, multiple copies are integrated into the same site. In a manner that is analogous to the techniques used in hybridoma selection, the experimenter selects cell lines that express the protein of interest at the highest level. Once selected, these cell lines can be maintained as stable lines and grown in culture in large volumes indefinitely.

The use of animal cells as a host is usually reserved for the second phase of a genetic engineering experiment with the first transformations carried out in bacteria. Once the sequence of interest has been identified it can then be transferred to the animal cell for other investigations.

Other alternatives for the expression of genes of interest also exist. These include plant cells, transgenic animals, transgenic plants and insect cell line-based baculovirus virus systems and fungi such as *Aspergillus* and *Fusarium*. The choice between these reflects both the skills and equipment available to the researcher and the end points which are required in a project.

5.2.3 Transgenic animals

Work by a number of research groups lifted genetic engineering to a new and controversial plane when it was shown that the genetic composition of an animal could be altered in a way that gave rise to inheritable characteristics. The most dramatic example of this came in 1982 when a mouse that was twice normal size was generated. This mouse was termed 'transgenic' as it contained the gene coding for growth hormone from rat stably integrated into the genome of its cells. The mouse was made by taking a fertilized mouse egg and micro-injecting the rat growth hormone gene through the egg wall and into the pro-nucleus. This impregnated egg was then transferred to a suitably prepared recipient female mouse and after a normal gestation period transgenic mice were born. Since then the production of transgenic mice has become widespread proving a very useful tool in studying gene expression. Transgenic mice that may prove valuable in identifying carcinogenic substances or agents have been patented, opening a public debate on the ethics of patenting organisms. The efficiency is approximately 30% of offspring being transgenic in the first generation with a normal inheritance of the trait subsequently. Specialized equipment, including micromanipulators, is required. Mendelian inheritance requires that the transgene be incorporated into the germ cells. Theoretically, if the transgene integrates into the genome of the fertilized egg before the egg divides, then all cells — both germ and somatic — will be transgenic. However, the transgene may often persist in the nucleus for a period of time while the egg is dividing before integration with the result that only a proportion of the animal's cells are transgenic. Such animals are called mosaic and only a small proportion of their offspring will carry the transgenic trait. Problems with transgenic animals are that sometimes the animals produced are sterile or the introduction of new genes is lethal during the early development of the animal. Such phenomena clearly reduce the usefulness of using transgenic animals.

Similar technology has now been used to develop transgenic cattle, sheep, birds and fish (see *Figure 5.2*). The idea of using live animals as bioreactors to produce useful substances has gained credence with many companies and also produced much public debate. An example of such a research program is the production of blood clotting factors in the milk of transgenic sheep from where they can be easily purified. This is an alternative to the use of animal cells in culture and although the economics of the processes are currently difficult to quantify the possibilities of having the proteins prepared on a very large scale in a medium that is easy to collect is attractive.

5.2.4 Transgenic plants

Micro-injection, electroporation, cell bombardment with gold particles associated to DNA and a bacterial vector — *Agrobacterium* — have been used to introduce foreign DNA into plants to create transgenic species. The emphasis has been on producing plants with new characteristics to improve their agriculture, for example improved disease resistance and optimal growth, rather than their use as bioreactors

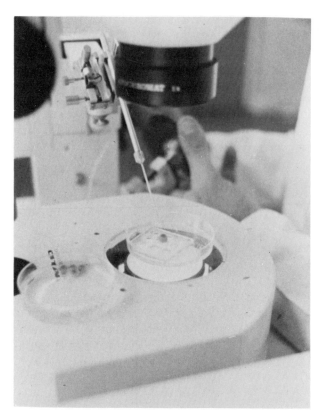

FIGURE 5.2: *Introduction of DNA into a salmon ovum by micro-injection. The foreign DNA or transgene is being transferred into the salmon ovum through a fine capillary glass tube. The operator is using micromanipulators delicately to puncture the ovum wall with the glass tube while viewing the operation through the microscope.*

for the production of other compounds. An example that has already reached the difficult stage of marketing is transgenic tomatoes with a longer shelf life. Environmental groups have raised the public consciousness of the issue of transgenics in terms of the escape of novel genetic material and possible as yet unknown hazards associated with modified plants. In response to the protests against the use of animals as bioreactors, some plants are also being used to produce animal proteins, for example the production of antibodies in potato plants is being actively investigated.

References

1. Griffith, F. (1928) *J. Hyg.,* **27,** 113.
2. Mandel, M. and Higa, A. (1970) *J. Mol. Biol.,* **53,** 159.
3. Cohen, S.N., Chang, A.C.Y. and Hsu, L. (1972) *Proc. Natl Acad. Sci. USA,* **69,** 2110.
4. Hanahan, D. (1983) *J. Mol. Biol.,* **166,** 557.
5. Murray, N.E. and Murray, K. (1974) *Nature,* **251,** 476.
6. Herskowitz, I. (1973) *Annu. Rev. Genet.,* **7,** 289.
7. Sternberg, N., Tiemeier, D. and Enquist, L. (1977) *Gene,* **1,** 255.
8. Scalenghe, F., Turco, E., Edstrom, J.E., Pirrotta, V. and Melli, M. (1981) *Chromosoma,* **82,** 205.
9. Dower, W.J., Miller, J.F. and Ragsdale, C.W. (1988) *Nucleic Acids Res.,* **16,** 6127.
10. Fiedler, S. and Wirth, R. (1988) *Anal. Biochem.,* **170,** 38.
11. Morgan, S.J. and Darling, D.C. (1993) *Animal Cell Culture.* BIOS Scientific Publishers, Oxford.

6 Screening of Clones in Libraries

6.1 Introduction

The techniques available for generating a variety of cDNA or genomic DNA libraries have been covered in previous chapters (see Chapters 3 and 4). The identification of a specific clone within a DNA library is the subject of this chapter. The goal is to identify, by some visual system, an individual colony or plaque (clone) amongst the myriad of others. At this stage the reader should visualize a library consisting of thousands of colonies or plaques each containing unique sequences inserted in an appropriate vector growing individually on agar plates. Screening a library is the process by which research scientists find the clone of interest. There are three major options for screening such DNA libraries: these are nucleic acid, phenotypic and immunoscreening. Successful nucleic acid screening depends on the target DNA sequence being present in a colony and available for hybridization (binding) with a complementary nucleic acid probe sequence (either DNA or RNA). A probe is a nucleic acid sequence (or group of sequences) which is chemically modified to allow its detection when bound to a complementary nucleic acid sequence. We will discuss probes in more detail later. Nucleic acid hybridization is by far the most favored method.

Phenotypic screening requires that the target clone expresses a protein product that is detectable by a plate assay of some type. For example, if one were screening for an enzyme activity such as a protease one could incorporate the substrate, a protein source such as skimmed milk, into the agar plate and visualize the clone of interest by a clearing around the colony. Immunoscreening requires that the target product be expressed in a clone and made available for binding to antibodies directed against the protein product.

The choice of screening method is made prior to generation of the library and depends largely on the clone being targeted and the type of library which is being generated. All three options for screening may be available but it must be stressed that this is rarely the case. The choice of screening method depends largely on the information available concerning the target clone. For instance, if the DNA sequence of the target is known, nucleic acid screening is generally the easiest option. However, if no sequence information is available the other options may be more suitable. For example, if the target protein has a known activity which can be detected by incorporation of a substrate in the agar plate, it may be possible to detect this activity in the vicinity of a specific clone. Alternatively, if there is an antibody available that binds to the target protein, this can be used to pinpoint the colony of interest. The relative merits of each method will be discussed in more detail in the following sections.

There is also the consideration of expense and time when choosing a screening method. In general, phenotypic screening is the least expensive as there are few costly manipulations after the agar plate stage. Nucleic acid screening is considerably more costly requiring expensive membranes and labeled probes (usually radiolabeled) although there is now an increasing move towards the use of nonradioactive screening techniques. It is also time consuming. Immunoscreening also requires membranes and expensive antibody labeling techniques, and can take a lot of time, particularly if the antibody to be used is not available and needs to be produced in a rabbit. The different screening methods will now be considered in more detail.

6.2 Nucleic acid screening

By far the most frequently used method is nucleic acid screening as this does not have some of the limitations of the other two methods. The major reason for its popularity is that this method does not require the target DNA to be expressed; it simply requires that the nucleic acid probe be able to hybridize to a complementary target sequence. Another major advantage is the versatility of probe design, the choice of probe being determined by the availability of nucleic acid material, the target being probed and the sequence information available. Before considering the types of probes available and the context in which they are employed, it would be useful to consider the principles involved in hybridization experiments as these are crucial to understanding probing strategies and results.

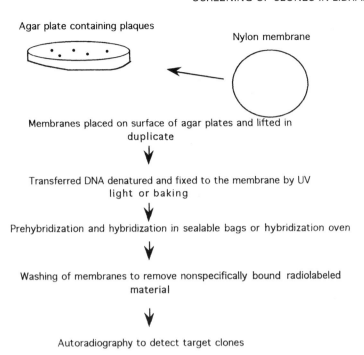

Agar plate containing plaques

Nylon membrane

Membranes placed on surface of agar plates and lifted in
duplicate

Transferred DNA denatured and fixed to the membrane by UV
light or baking

Prehybridization and hybridization in sealable bags or hybridization oven

Washing of membranes to remove nonspecifically bound radiolabeled
material

Autoradiography to detect target clones

FIGURE 6.1:   *Flow chart of screening procedure using radiolabeled
probes. In the case of a phage library the nylon membranes can be placed
directly on to the plates and lifted off leaving double-stranded DNA
attached. Denaturation, fixing, hybridization and autoradiography follow
as outlined.*

In Chapter 1 the concept of the double-stranded nature of DNA and
the complementarity of DNA sequences was discussed. This is the
fundamental property of DNA that is exploited when using nucleic
acid screening techniques. The manner in which pieces of DNA can be
joined up to create structures that can be propagated in bacterial cells
has also been discussed. How can these be analyzed such that a particular DNA sequence of interest can be identified? The methods are
described in detail in ref. 1 and a schematic diagram is shown in
Figure 6.1. The first step is to separate each sequence, which is
achieved by transforming bacteria with ligated vector and target
DNA, and thereby generating individual bacterial colonies. The next
step is to access the DNA in the individual cells. This begins by transfer of the colonies or plaques in duplicate on to nylon membranes by
placing fitted membranes on to the agar plate containing the target
organisms. The membranes are then peeled off and the bacteria lysed
on the membrane in a buffer solution. After lysis and washing, the

exposed DNA is denatured with alkali treatment followed by neutralization. This is the most crucial step in the procedure as it separates the DNA strands making them single stranded and therefore available for hybridization with complementary labeled DNA (i.e. nucleic acid probe). The filters are then either treated with ultraviolet (UV) light or baked in order to fix the separated DNA strands to the membrane.

The next stage of the procedure is to 'probe' the bound DNA with chosen complementary DNA sequences. The ability of complementary DNA strands to hybridize to each other, under well-established experimental conditions in solution, is exploited. As mentioned earlier, nucleic acid probes can take many forms although, for the purpose of screening DNA libraries for a specific sequence, they are generally of two types: (i) double-stranded DNA probes and (ii) oligonucleotide probes. The theory of hybridization is complex and beyond the scope of this text; however, there are excellent books written on the subject and the reader is referred to these (see Appendix B).

The membrane (filter) is then exposed to the DNA probe for hybridization analysis. This is usually performed in a sealed plastic bag. The first step is to prehybridize the filters in a pre-prepared salt solution containing a variety of high molecular weight molecules which 'block' the filter. This is to prevent nonspecific binding of the probe mix to the membrane (sample), thus preparing it for the subsequent hybridization steps. Following prehybridization the solution is removed and the hybridization solution containing the labeled probe is added to the bag containing the filters. The process of hybridization under carefully chosen conditions allows the labeled probe to bind to target sequences. However, after hybridization there can be a high degree of nonspecific binding, which it is important to wash away. The conditions of hybridization although enabling the binding of probe sequences to complementary DNA do not prevent them from binding to sequences which are not complementary. In most probe hybridization experiments it is necessary to wash off probe that is nonspecifically bound. This is done by increasing the stringency of hybridization in the washing step by increasing the temperature of incubation and/or decreasing the salt concentration in the wash solutions. Both approaches will allow only the probe sequences with absolute complementarity to be retained on the filter hybridized to the target DNA. Thus, by altering these parameters, one can reduce background nonspecific binding. Having washed the filters at the appropriate stringency they are then placed in sealable bags. At this stage of the procedure the aim is to have a numbers of filters with discrete areas of bound DNA derived from plaques or colonies, some of which have a bound complementary probe which will allow visual detection.

How is visual detection achieved? There are a number of different ways in which a probe can be labeled in order that it emits a visually detectable signal. The most popular and sensitive method is the use of radioactivity. Radiolabeled nucleotides can be incorporated into nucleic acid sequences by a number of different enzymatic reactions. In the case of double-stranded DNA, radioactive nucleotides are incorporated into nucleic acid material using DNA polymerase (see Chapter 2). Traditionally, a technique called nick translation was used; however, this has largely been superseded by a technique termed multiprime labeling in which short oligonucleotides are used to prime extension by DNA polymerase from the target DNA being used for probing. In the case of oligonucleotide probes, a method called end labeling is used in which T4 polynucleotide kinase is used to label the 5'-hydroxyl end of the nucleotide (see Chapters 1 and 2). Once radiolabeled, the 'hot' nucleic acid probe emits a signal that is detectable on X-ray film in the form of a dark spot. This process is termed autoradiography (*Figure 6.2*).

Therefore, the final step in nucleic acid screening is to place the filters on a grid inside a cassette. A sheet of X-ray film is then placed over the filters which are usually covered with a film of plastic. The cassette is closed and placed in a freezer for up to 2 weeks. The film is then removed and developed. If the radiolabeled probe has bound to a discrete sequence it will cause a black dot to form on the film at a precise location. This position can be matched with the original agar plates from which colonies or plaques have been lifted (remember the first step in screening). This is the process by which we screen for a specific sequence contained in the clone. Having identified this colony or plaque the next step is to isolate and enrich the clone and to analyze it further. This is the subject of the next chapter. The process described above is a simplification of the methods involved in nucleic acid screening.

Recently nonradioactive methods of labeling probes in a manner similar to that outlined above have been developed. These involve the use of labels such as digoxygenin and biotin. However, although these methods are extremely sensitive and have the advantage of safety, they have not gained universal appeal in the molecular biological community [2].

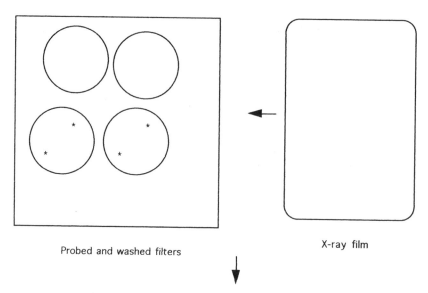

Probed and washed filters   X-ray film

X-ray film is placed over the probed membranes and placed
in a cassette to allow signal development

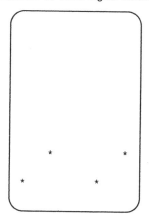

FIGURE 6.2:   *Autoradiographic detection of probes bound to target sequences. In the top left corner, four membranes (represented by circles) which have gone through the hybridization process are shown. The bottom two are duplicates and sites where a probe has bound are indicated by asterisks. In order to visualize where the radiolabeled probe has bound, an X-ray film is placed on top of the membranes. A signal develops on the X-ray film at the precise location where the probe is bound. This is shown in the bottom part of the figure. The signal can be traced back to a location on the original agar plate, and the appropriate clone can be targeted for further analysis.*

6.3 Phenotypic screening

This method for screening a DNA library is used infrequently for a number of reasons. Firstly, it requires that the clone of interest expresses an activity that is easily detectable on an agar plate. Therefore, it is necessary that the entire gene or the active portion of the gene encoding the activity be present in the cloned DNA. In general, the entire gene is needed. It also requires that the DNA is transcribed and translated. For these reasons, this screening method is usually confined to cloning genes from organisms which have transcription and translation systems that are compatible with that of the host. Therefore, it can only be used in limited circumstances, which is a major drawback.

When utilizable the method has the advantage that it is extremely rapid and inexpensive compared with other methods. There are many examples of the use of this method available in the literature, the most frequent being the cloning of extracellular enzymes such as amylases and proteases from *Bacillus subtilis* [3, 4]. An example of this technique is shown in *Figure 6.3*.

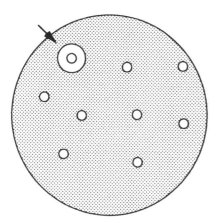

FIGURE 6.3: *Schematic of phenotypic screening. Bacterial colonies are growing on an agar plate in the presence of a substrate for the target cloned gene. If the gene is expressed in one colony, and the protein product is exported out of the cells, it can degrade the substrate creating a 'halo' effect around the colony. This is indicated by an arrow.*

6.4 Immunoscreening techniques

Immunoscreening techniques are used when there is no DNA sequence information available for the gene of interest but an antibody is available. In this case, the library must be made with an expression vector. There is a requirement for the protein product of the inserted DNA to be present in the colony or plaque. The procedure for making the protein accessible to antibody binding is similar to that of nucleic acid screening. The colonies or plaques are lifted on to nylon membranes in duplicate from the agar plates. In the case of colony lifts, the cells need to be lysed to make the target protein accessible to the antibody. In the case of plaques, the bacteria have lysed so the protein is available for binding. The membranes are put through washing steps to ensure that the best possible conditions are available for the binding of the antibody. Detection of bound antibody can be achieved by the use of radiolabeling followed by autoradiography, as described previously [5].

References

1. Maniatis, T., Fritsch, E.F. and Sambrook, J. (1989) *Molecular Cloning*, 2nd Edn. Cold Spring Harbor Laboratory Press, Cold Spring Harbor, NY.
2. Jones, P., Qiu, J. and Rickwood, D. (1994) *RNA Isolation and Analysis*. BIOS Scientific Publishers, Oxford.
3. Ortlepp, S.A., Ollington, J.F. and McConnell, D.J. (1983) *Gene*, **23**, 267.
4. Honjo, M., Manabe, K., Shimada, H., Nakayama, A., Mita, I. and Furutani, Y. (1984) *J. Biotechnol.*, **1**, 265.
5. Broome, S. and Gilbert, W. (1978) *Proc. Natl Acad. Sci. USA*, **75**, 492.

7 Analyses of Selected Clones

Once the individual clones of interest have been selected from a gene library, genetic engineers may congratulate themselves on the success of obtaining a rapid source of significantly greater amounts of genetic material for study than is possible by standard biochemical methods. A choice must then be made as to which methods for the analysis of the identified recombinant DNA clone are appropriate for the overall aim of the project. Questions often asked include: is the elucidation of the DNA sequence of the recombinant DNA required? Is analysis of the genomic locus of the recombinant DNA required? Can analysis of transcription of the cloned gene be performed? Has the selected DNA fragment the capability to be translated into protein and if so is the study or purification of the protein product important? Most likely, the answers to all or a combination of these questions will be required. Therefore a variety of options are available, the most common of which are explained below.

7.1 DNA or genome analysis

Methods of analysis of the isolated DNA allow one to estimate the length of foreign DNA ligated to vectors and selected in individual clones and eventually to elucidate the nucleotide sequence of the cloned DNA. Several procedures exist for each manipulation and every laboratory will have its particular favorite (or often, each researcher in a laboratory!). However, they all share common features in design, primarily, the ability to prepare sufficient quantities of pure-quality DNA for analysis in a relatively short time period. The abundance of specialized 'kits' from manufacturers of molecular biology reagents also gives the researcher a multitude of choices in selecting a suitable method for a particular form of analysis.

7.1.1 Rapid preparation of recombinant DNA from clones

Rapid preparation of recombinant DNA from the bacteria selected is normally the first analysis performed on clones that have been isolated from gene libraries. The researcher will probably be interested in rapidly confirming that the isolated clone contains the foreign DNA of interest and also will wish to estimate the length of the foreign DNA fragment cloned. This can quickly establish a level of confidence that a complete gene has been cloned or indicate that only a portion of a gene has been obtained and that further screening of gene libraries is required. Also, many clones may have been identified as 'positives' after screening gene libraries and the investigator will wish to see which has the largest insert or if any clones share similarities in size or in restriction enzyme digestion patterns, that is whether they are probably derived from the same fragments of DNA of the genome. Therefore, methods for rapid preparation of recombinant DNA from clones have two features: the first being speed (1–2 days to perform) and the second being the ability to isolate DNA which is sufficiently pure (no contamination with cell components such as proteins, lipids or carbohydrates) for analysis. The analysis will usually consist of restriction enzyme digestion and gel electrophoresis. Usually, 1–10 ml of nutrient broth are inoculated with a positive colony or plaque and incubated overnight. Cells containing recombinant plasmids or lambda phage containing recombinant DNA are then collected from the broth — usually by centrifugation. The bacterial or phage membranes and walls are disrupted by a combination of heat treatment, digestion with proteinases and treatment with detergents. This releases the cloned DNA which is then selectively precipitated in relatively pure form from the cell or phage lysates by the addition of ethanol. High-speed centrifugation of the ethanol solution will pellet the DNA which, after removal of the ethanol, can be resuspended in sterile water and stored at −20°C prior to further analysis. All rapid DNA preparations for recombinant plasmid or phage DNA involving *E. coli* cells share the features of 1–10 ml overnight cultures to produce sufficient *E. coli* biomass followed by rapid DNA isolation [1,2]. This makes it possible to analyze 10–20 clones of interest producing 5–10 μg of recombinant DNA. This will be sufficient for 5–10 restriction enzyme digestions, allowing the researcher to advance the analysis of the isolated DNA in a short time period.

7.1.2 Southern blot analysis

One of the most frequently cited publications in molecular biology is the paper by E.M. Southern [3] describing the ability to transfer DNA from agarose gels on to solid supports (e.g. nitrocellulose membrane sheets or filters). These filters can then be analyzed by probing with a particular DNA fragment to determine whether DNA fragments of similar nucleotide sequence are bound to the filter. The procedure, termed Southern blotting after the author, is one of the most routine and common techniques in all genetic engineering laboratories. This reflects its usefulness to the scientist for gathering information. It allows one to identify which DNA fragments, resolved by gel electrophoresis, are recognized by any particular DNA molecule of interest. The DNA molecule of interest is in this case called a 'DNA probe' and it is tagged with a specific label (often a radioactive isotope such as ^{32}P, see Chapter 6). The method allows this DNA probe to find and interact, by base pairing of complementary regions of nucleotides, with any similar DNA molecules bound on the filter. Exposure to an X-ray film (autoradiography) will then result in a band (or bands) becoming apparent on the film which can then be correlated with photographs of the original gel to determine DNA fragment sizes.

The basic steps of the procedure are as follows. DNA fragments produced as a result of restriction enzyme digestion are separated by agarose gel electrophoresis. Gels are stained with ethidium bromide, visualized over a UV source and photographed. A ruler is placed alongside the gel for photography to become a reference to determine the lengths of DNA fragments identified by the DNA probe after the Southern blot. The gel is then allowed to soak in a denaturing solution, usually 0.5 M NaOH. The sodium hydroxide will disrupt the hydrogen bonding of the double-stranded DNA molecules. This will, at a later stage, allow the denatured strands to associate by hydrogen bonding with the DNA probe. The gel is then soaked in neutralizing solution which corrects the pH to neutral. The DNA fragments are now denatured *in situ* in the gel. They are then transferred to a solid support — originally nitrocellulose filters but nylon filters are now becoming more fashionable. This transfer is termed blotting and the absorption of salt solution through the gel and filter by the paper napkins results in the transfer of DNA from the gel on to the filter. The transferred DNA fragments are then irreversibly immobilized in a single-stranded state by treating the filter for 2 h at 80°C (or an exposure of 2–5 min with UV light for nylon filters). An exact copy of the gel pattern of DNA fragments on the nitrocellulose or nylon membrane is thereby obtained. This membrane can then be soaked in a

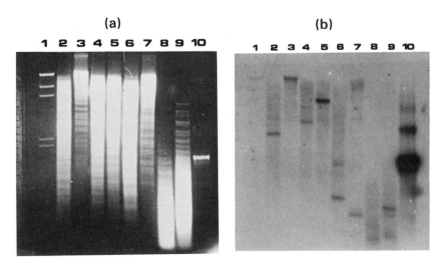

FIGURE 7.1:   *Restriction digestion followed by Southern blot and hybridization analysis of genomic DNA. The photograph in (**a**) represents the results of the digestion of bacterial genomic DNA with a variety of restriction nucleases. Lane 1 contains DNA fragments of known lengths. Lanes 2–9 represent bacterial genomic DNA digested with EcoRI, KpnI, HindIII, PstI, BamHI, SacI, HaeIII and Sau3A, respectively. Lane 10 contains a sample of a 1.8-kb DNA fragment which will be used as a DNA probe to screen a Southern blot of this agarose gel. Note, the smaller size range of fragments produced by the 4-cutters HaeIII and Sau3A in comparison with the 6-cutter enzymes. The photograph in (**b**) represents the autoradiograph after Southern blot and hybridization of the above gel with the 1.8-kb DNA probe. Note, the very strong hybridization of the DNA probe to itself in lane 10. Individual DNA fragments hybridizing to the DNA probe can be detected in most of the restriction digestions.*

solution containing a tagged DNA probe. After a period of hours, to allow the DNA probe to find and interact with similar DNA fragments, this hybridization is halted, the filter washed to remove the solution and finally exposed to autoradiograph film. The resulting hybridizing DNA fragments on the filter can be analyzed by comparison of electrophoretic distance using a ruler and correlation with the photographed gel (see *Figure 7.1*).

7.1.3 DNA sequencing

The determination of the nucleotide sequence of a DNA fragment is often a critical step in its analysis. DNA fragments are resolved by gel electrophoresis and exposed to autoradiograph film which produces the information for analysis. However, in this case, high-resolution denaturing polyacrylamide gels are used. These can resolve DNA fragments differing by only 1 nt in length. Four reactions are performed in parallel designed to identify the location of each A, G, C or T in the DNA. These fragments must have a radioactive tag incorporated to allow their visualization. Practically, 300–500 nt of sequence can be analyzed in one reaction using these electrophoresis systems. Two procedures are used for DNA sequencing — the dideoxy or enzymatic chain termination method developed by Sanger in 1977 [4] (see *Figure 7.2*) or the chemical method developed by Maxam and Gilbert [5], also in 1977 (see *Figure 7.3*). The dideoxy method is probably the most common in application. The products of the sequencing reactions are resolved and, after autoradiography, the DNA sequence can be read directly from the film (*Figure 7.4*).

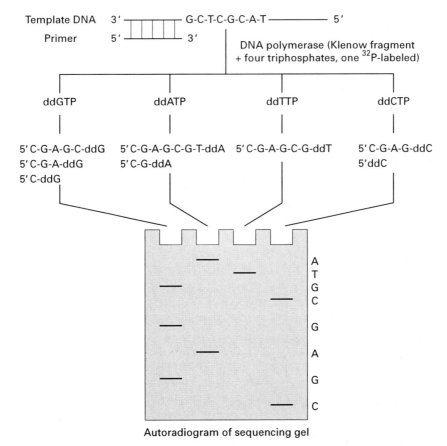

Autoradiogram of sequencing gel

FIGURE 7.2: *DNA sequence analysis by the dideoxy chain termination (Sanger) method. Reproduced from ref. 6.*

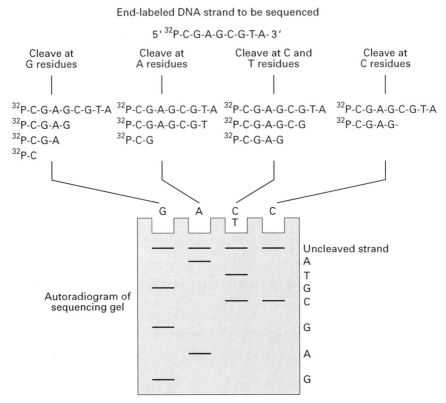

FIGURE 7.3: DNA sequence analysis by the chemical degradation
(Maxam and Gilbert) technique. Reproduced from ref. 6.

The chemical method uses agents that cleave DNA at specific bases.
Sequencing products are resolved and autoradiographed in a similar
fashion to the dideoxy method.

7.2 RNA or transcript analysis

In a similar fashion to DNA, RNA can be prepared and analyzed by
gel electrophoresis and subsequent hybridization and autoradiography [7]. This allows the investigator to determine whether a gene is
transcribed in a particular cell type as opposed to others or whether
it is expressed at different levels during different stages of the cell life
cycle. This is relevant when studying tissue-specific expression or
developmental expression of genes. Similar to DNA analysis, it is also
possible to estimate RNA length with this form of analysis.

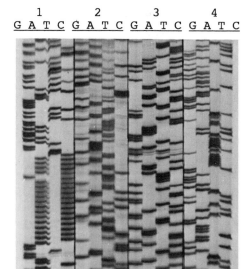

FIGURE 7.4:   *DNA sequencing. Photograph of a portion of an autoradiograph of a DNA sequence gel. Eacch clone (1–4) is represented by four lanes (G, A, T, C) corresponding to each nucleotide. The DNA sequence of a clone can be determined by reading up the gel noting the next closest band from any of the four lanes. For example, one can clearly see and read the ACACACAC repeat sequence in clone 1.*

Following on the nomenclature of the Southern blot, the transfer of RNA from agarose gels on to solid support membranes (again nitrocellulose or nylon) is termed Northern blotting. The main variation in the technique reflects a characteristic difference in the structure of RNA and DNA, that is DNA is double stranded and RNA is single stranded. Therefore, RNA gels do not have to be soaked in denaturation and neutralization solutions before transfer. However, the gel systems are also different. Single-stranded RNA has a tendency to form secondary structures by the base pairing of short stretches of complementary ribonucleotides along its length. This secondary structure affects RNA migration in gels so that true lengths of RNA molecules cannot be elucidated unless this secondary structure is disrupted. Usually, a denaturant is added to the gel solution and this ensures that any hydrogen bonding is broken and the RNA molecules migrate as true single-stranded molecules. The addition of formamide as the denaturant and the heating of the RNA preparation before

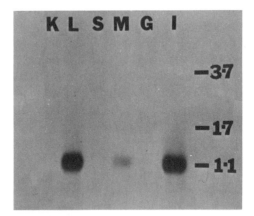

FIGURE 7.5:   *Northern blot and hybridization. The photograph represents an autoradiograph of a Northern blot and hybridization analysis. Lanes K to I represent RNA prepared from kidney, liver, spleen, muscle, gonad and intestine tissue, respectively. The labels 3.7 and 1.7 are known RNA lengths (kb) marking the distance migrated by the 28S and 18S ribosomal RNA fractions in the various tissues. A positively hybridizing mRNA of 1.1 kb in length is seen to be present in both liver and intestine RNA (lanes L and I) and also at a lower level in muscle (lane M).*

electrophoresis ensure single strandedness. These denaturing gels are then electrophoresed in the same fashion as normal agarose gels and directly blotted using the Southern methodology. Hybridization with radioactive DNA probes followed by autoradiography is performed in a similar manner to that for the Southern blot (*Figure 7.5*).

7.3 Protein or translation analysis

Just like DNA and RNA, proteins can be separated and analyzed by gel electrophoretic systems [8]. This can be important following the isolation of a DNA fragment of interest, particularly if one wishes to study the protein it encodes. Denaturing SDS–polyacrylamide gels are used for such analyses. Proteins of different sizes are fractionated and molecular weights can be determined by comparison with standards resolved on similar gels. Total protein preparations from either prokaryotic or eukaryotic cells will result in a multitude of protein bands when resolved by electrophoresis. However, the protein of interest can be identified from the multitude by a method analagous to Southern and Northern blots. Maintaining the convention, this method is termed the Western blot.

7.3.1 Western blot analysis

Proteins resolved by polyacrylamide gel electrophoresis are transferred to a nitrocellulose membrane. This transfer, unlike with DNA and RNA, requires an electric charge to transfer the proteins from the gel on to the membrane. This electroblotting is performed by placing the gel and membrane in an electric chamber submerged in electrophoresis buffer and applying an electric current for a period of hours. Afterwards, this membrane can then be screened with a 'protein probe' specific for the protein of interest. This protein probe is tagged with a radioactive label (usually ^{125}I). Typically, the protein probe is an antibody for the protein of interest and so will recognize its specific protein antigen on the membrane, bind to it and, following subsequent washing and autoradiography, a band corresponding to a protein on the original gel will be apparent. In this case, one of the major experimental considerations is the preparation of antigen to raise the appropriate antibodies. This can take time, for example several weeks to raise appropriate anti-sera in animals after injection of antigen. However, very many antibodies are commercially available and so quite often it is not necessary to prepare antibodies.

7.4 Beyond sequences

The previous sections have illustrated how the analysis of an isolated gene was initiated. The data provided will be the basis for all future work which increasingly will move the researcher from the study of the structural aspects of the gene to studies of how it functions. The range of methods used to carry out this analysis are expanding very rapidly and the criteria for acceptance of data that prove the functional relevance of a sequence are becoming extremely demanding. If one considers the common situation where a promoter region from the eukaryotic gene has been isolated and sequenced, the next step would probably involve scanning databases to see whether sections of the sequence have been implicated previously in some function. An example of this would be the occurrence of a target site for some known transcription activation factor in a region of the DNA under study which is thought to be a promoter. The target sites at which these factors bind and which are essential for the transcription of the gene are known to have some potentially variable sequences. It is therefore rare that an absolutely correct target sequence would be the consequence of such database scans. Rather an indication would be obtained that a candidate sequence for a transcriptional activation factor is present close to or at the promoter region of the gene of interest.

If there are such transcriptional activation factors that influence the functioning of the gene then it follows that these should bind to the DNA target sequence. There are methods available that illustrate that such a binding has occurred. In the first of these the putative target region is chemically synthesized, radioactively labeled by end-labeling, incubated with a tissue homogenate which is liable to contain the transcription activation factor and subsequently subjected to electrophoresis. If the target oligonucleotide has not interacted with a protein in a specific manner then it will electrophorese at a size which is dependent on its own molecular weight. If, however, it has formed a strong linkage with another protein then the oligonucleotide will be retarded during electrophoresis and will give rise to a new band at an apparently higher molecular weight. If such a result is obtained one might think that the linkage between the target sequence and the role of the transcriptional factor has been established. Of course, this is not necessarily the case. One variation on the experiment that would consolidate the result would be the use of mutated targeted sequences to show that such an alteration of the target sequence results in a loss of the protein binding. However, this only shows that a protein, as yet undefined, binds to that particular target region. In some cases it is possible to confirm directly that the retardation of the oligonucleotide during electrophoresis was due to a particular protein by use of an antibody directed against that protein. This changes the molecular weight of the complex and hence by a 'supershift' gives rise to a new band after autoradiography. A complementary series of experiments will place these results in a wider context. In this case, a DNA region of several hundred nucleotides which encompasses the apparently interesting regions of the promoter is again end-labeled and incubated with a mixture of proteins from a cell extract or tissue homogenate. The DNA fragment is then subjected to DNase digestion in the presence and absence of the proteins. Under the appropriate conditions, a complete ladder of digestion is obtained which will show, in a manner analogous to DNA sequencing, a series of bands each of 1 nt side difference. The regions which have been protected from the DNase action by the binding of the protein will not yield a binding pattern at that location. This will indicate that binding has occurred, which could be seen as further evidence for a role of that sequence in the control of the expression of the gene.

Both of these approaches have the disadvantage that they are completely *in vitro* reactions. The relative concentrations of proteins and

DNA are artificial, as are the ionic conditions and consequently the possibility of artifacts cannot be excluded.

The next phase in the definition of the action of the promoter region could be transfection of the promoter region into recipient cells. Typically, the putative promoter region is linked to a reporter gene element, transfected into a cell line and the activity of the reporter gene subsequently monitored. The steps involved in this are similar to those outlined for the use of animal cells. The reporter genes currently in use include β-galactosidase, luciferase or chloramphenicol acetyltransferase (CAT). Following transfer of the test construct into the cell line the experimenter has the option of collecting the cells and analyzing the activity of the reporter gene while the vector is still in an episomal location. This is known as a transient transfection assay. It has the advantage of giving rapid results and can be used on a large number of samples. Alternatively, the experiment may be allowed to proceed until permanently transformed cell lines are obtained in which the vector is integrated into the chromosome of the animal cell in a random manner and the activity of the reporter gene is subsequently analyzed. This has the advantage of allowing further cell biology experiments to be carried out on the cell to indicate how the gene construct responds to different experimental conditions. There are obvious variations to this experiment which can be used to define the promoter region in more detail. One of these is the preparation of deletion or point mutations in the promoter region to highlight the role of different aspects of the promoter. Another is the use of different host cell lines which are known to contain or lack different transcription activators. By a combination of these experiments it is possible to show that the promoter region is functional and that certain sequences in it are important in the expression of the gene.

More recently, these experiments have been seen as a mere start to the process of true analysis of the functionality of the gene or analysis of the factors that act on it. The possibility of making transgenic animals, and particularly transgenic mice, leads to the possibility of showing that the functions attributed to the sequences can in fact work *in vivo*.

In this case, the impact of a whole range of variants of cell type, stage of development, sex, etc., can be analyzed in greater detail. In some cases, the gene that is implied to have a role in a particular cell or tissue will be further tested to prove this role *in vivo* by generating 'knock out' or dominant-negative mutant mice. In the 'knock out experiment' the gene of interest is inactivated by targeted integration of sequences that destroy the functional activity of the gene product. The impact of this procedure on the transgenic mouse is then monitored and a more complete idea of the role of the gene is established. In the dominant-negative experiment, a variant and inactive form of the gene of interest is overexpressed in an inducible manner, thereby occupying all target sites for the intact foreign protein and the consequences of this on gene expression are analyzed in the whole animal.

While the move towards whole animal experiments is developing, there is an equally strong move towards greater *in vitro* refinement. Having indicated that a particular protein is involved in binding to a target sequence and consequently has an impact on its expression, it is now increasingly possible and desirable to provide greater detail on the way in which this occurs. This involves the study of the protein–protein interactions that accompany the protein–DNA interactions. All of these studies are now moving to the use of X-ray crystallography and nuclear magnetic resonance (NMR) techniques to define the actual amino acids that are involved in either of these processes. Again, the proof that these interactions occur *in vivo* is increasingly a requirement with the consequent necessity of preparing a series of mutations in the primary structure of the protein of interest to demonstrate *in vivo* that the X-ray crystallography results have a real functional relevance.

Developments of the sort that are outlined above are of course to be applauded as the most sophisticated and complete methods of obtaining irrefutable evidence of the way in which genes function. The consequences in terms of the quality of publications in the area of molecular biology are equally and obviously dramatic. However, the costs of such experiments in terms of time, manpower required and money are substantial.

References

1. Birmboim, H.C. and Doly, J. (1979) *Nucleic Acids Res.,* **7,** 1513.
2. Leder, P., Tiemeier, D. and Enquist, L. (1977) *Science,* **196,** 175.
3. Southern, E.M. (1975) *J. Mol. Biol.,* **98,** 503.
4. Sanger, F., Nicklen, S. and Coulson, A.R. (1977) *Proc. Natl Acad. Sci. USA,* **74,** 5463.
5. Maxam, A.M. and Gilbert, W. (1977) *Proc. Natl Acad. Sci. USA,* **74,** 560.
6. Williams, J., Ceccarelli, A. and Spurr, N. (1993) *Genetic Engineering.* BIOS Scientific Publishers, Oxford.
7. Thomas, P.S. (1980) *Proc. Natl Acad. Sci. USA,* **77,** 5201.
8. Towbin, H., Staehelin, T. and Gordon, J. (1979) *Proc. Natl Acad. Sci. USA,* **76,** 4350.

8 The Polymerase Chain Reaction

8.1 Introduction

Genetic engineering has allowed scientists to study genes and how they work in a manner scarcely thought possible prior to the advent of the present technology. The capacity to search for a DNA sequence of interest and obtain enough for further study by gene cloning techniques has been of fundamental importance in our understanding of molecular biology. It was difficult to anticipate that any revolutionary technology would emerge that would make our studies even easier to perform. However, such a technology did emerge and was termed the polymerase chain reaction, known more popularly as PCR. In 1987 an article was published describing this technology [1]. The method allows amplification of a target DNA to sufficient quantities for further study using standard DNA analytical procedures, such as agarose gel electrophoresis, hybridization probing and sequencing. The PCR amplification is performed in a small plastic sealable tube *in vitro* as distinct from amplification of the sequence in a specialized bacterial or yeast vector *in vivo* using gene cloning techniques. The impact of the technology was immediate and revolutionary and it is now a well established technique in the field of molecular biology.

The reasons for this major impact are many. The method is extremely rapid; it takes only 3 h to amplify a known sequence of interest compared with 'classic' genetic engineering techniques which could take up to a week or more. The method is simple, as the PCR can be performed in a single tube with minimal components which are just dispensed together. Other gene cloning methods typically require expensive materials such as membranes and radiolabeled nucleotide triphosphates and specialized skills. The PCR can be performed on relatively crude DNA-containing samples, for example untreated blood for forensic analysis. This is in contrast to the standard methods in gene manipulation which require that the DNA, both target and vector, be relatively pure. These factors have made the PCR an attractive alternative to classical methods for the amplification of specific sequences. However, it must be stressed that the overwhelming majority of applications of PCR technology require that knowledge of the DNA target sequence is available. This is one limitation of the technology that does not apply to established gene cloning techniques where no knowledge of the sequence is required. For this reason, PCR builds on the foundations of genetic engineering techniques rather than replaces them.

8.2 Basic principles

The PCR is an enzymatically mediated *in vitro* amplification of a specific target DNA sequence. This is achieved by simultaneous primer extension of complementary strands of DNA. This is generally obtained by repeated cycles (up to 35) of heating (95°C), cooling (37°–65°C) and extension at 72°C using a thermostable DNA polymerase. The reaction is carried out in a plastic tube inserted in a regulatable heating block termed a thermocycler (PCR machine) which can be programed to heat and cool in the sequence required. This results in exponential amplification of a target sequence (*Figure 8.1*). To the uninitiated, this description of the PCR may seem complicated and uninformative. However, all the basic information which was used to conceive the idea of PCR is available in the earlier sections of this book. In Chapter 1 the concept of DNA complementarity and the 5'–3' nature of DNA synthesis was introduced. In addition, the concept of primer extension was also covered while in Chapter 2 the properties of DNA polymerases were described.

Before discussing the components of a PCR reaction it seems appropriate to consider first the principle underlying the method (*Figure 8.1*). In the first cycle of the PCR the template DNA provides a target for the primers. Remember the process involves primer extension. As described earlier, DNA is synthesized in a 5'–3' direction. For the purposes of illustration the molecular biologists' short hand for DNA will be used as shown in *Figure 8.2*.

As PCR involves simultaneous primer extension of the complementary strands two primers, P1 and P2, both synthesized in a 5'–3' direc-

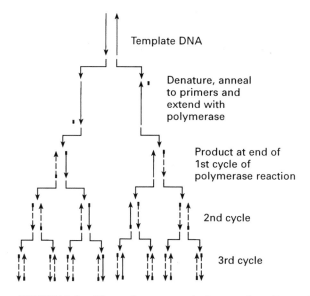

FIGURE 8.1: *The polymerase chain reaction. Reproduced from ref. 2.*

tion, are designed to be complementary to discrete sequences on the respective strands. When the strands are separated by heat and cooled the primers anneal to their target sequences. This is represented diagramatically in *Figure 8.3*.

In the presence of the thermostable DNA polymerase, buffer, MgCl$_2$ and the dNTPs, extension occurs from the bound primers resulting in the synthesis of new strands of DNA (*de novo* synthesis) in the 5'-3' direction (see *Figure 8.4a*). Thus, both strands are replicated from fixed start points defined by primers P1 and P2. These products can be described as first cycle primer extension products (*Figure 8.4b*). If these start points for extension by P1 and P2 are sufficiently close to each other synthesis proceeds past the site of binding of the other primer. Thus, the target for annealing of P1 is synthesized from P2 and vice versa (see *Figure 8.4b*).

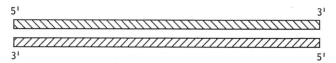

FIGURE 8.2: *Short hand for double-stranded DNA showing the 5' to 3' direction of the top strand and the 3' to 5' direction of the bottom strand. This represents the target DNA prior to PCR. By convention, top strand 5'–3' represents the coding strand and is usually the one presented in the literature as the DNA sequence.*

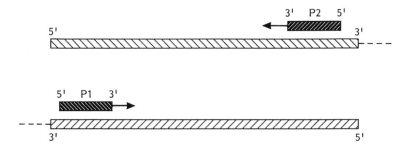

FIGURE 8.3: *Separation of the complementary strands of DNA allows primers P1 and P2 to bind to their complementary sequences in the target DNA. Note the 5'–3' directionality in all cases.*

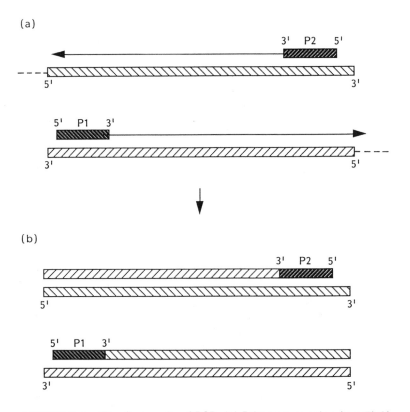

FIGURE 8.4: *The first cycle of PCR. (**a**) Primer extension in a 5'–3' direction of the target DNA from primers P1 and P2 in the presence of DNA polymerase, reaction buffer, deoxynucleotide triphosphates and MgCl₂. (**b**) New DNA strands synthesized generating new target sequences for primers P1 and P2.*

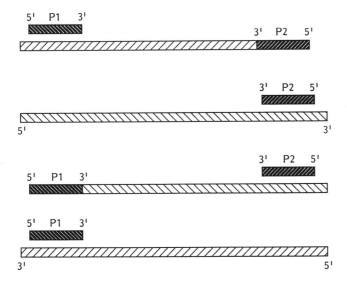

FIGURE 8.5: *Second cycle of PCR illustrating the separation of the strands from cycle 1 (*Figure 8.4*) and annealing of the primers to the newly synthesized and original targets.*

At the start of the second cycle, the double-stranded products of the first cycle are heated and separated into two single strands. All the resulting DNA molecules then act as templates for binding of primers P1 and P2 as shown in *Figure 8.5*. The annealed primers are extended by the thermostable polymerase as before and the resulting products are shown below in *Figure 8.6*.

It is worth noting that after the second cycle there is already an enrichment of sequences containing the P1 and P2 primers. Six out of eight of the PCR products are extended from these primers and two of them have lengths defined by P1 and P2. Thus, sequences between P1 and P2 are already being amplified. In the third cycle of PCR (see *Figure 8.7*) the products of the second cycle are templates for binding of primers 1 and 2. After completion of this cycle eight out of 16 products contain the P1 and P2 sites and intervening sequences. This particular molecule is exponentially amplified. The original primer-extended product (see cycle 1) is amplified in a linear fashion.

8.3 The reaction components

A PCR can be broken down into its individual components or starting material: the template which is the target for amplification, the

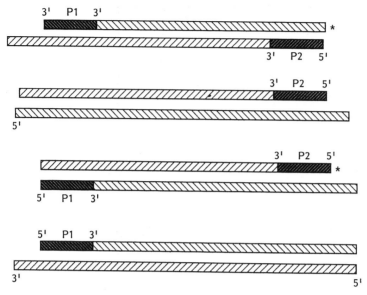

FIGURE 8.6: *Products generated after completion of the second cycle of PCR. * Indicates products with boundaries defined by P1 or P2 or their complementary sequences.*

oligonucleotide primers (referred to as amplimers, oligos or primers), dNTPs, and the thermostable DNA polymerase with an appropriate reaction buffer and $MgCl_2$. There are recommended concentrations for each of these components. These are available from suppliers and in many publications which consider experimental protocols to be important. As this text is concerned mainly with principles, further reading of a number of excellent texts is recommended (see Appendix B).

8.3.1 Thermostable polymerases

The utilization of thermostable DNA polymerases was critical to the development of PCR technology. Prior to its introduction, experimenters used the thermolabile 5'–3' DNA polymerase activity termed Klenow fragment. After each cycle of heating a fresh aliquot of the enzyme had to be added and, since each cycle usually takes about 3 min, the expense and practical difficulties made use of the Klenow fragment enzyme prohibitive. Then thermostable polymerases arrived on the scene. They are obtained from organisms that thrive in hot springs and have an optimum activity at 72°C. Thus, there was no need for extra additions of enzyme during the cycling process where strand separation required heating to over 90°C and extension could

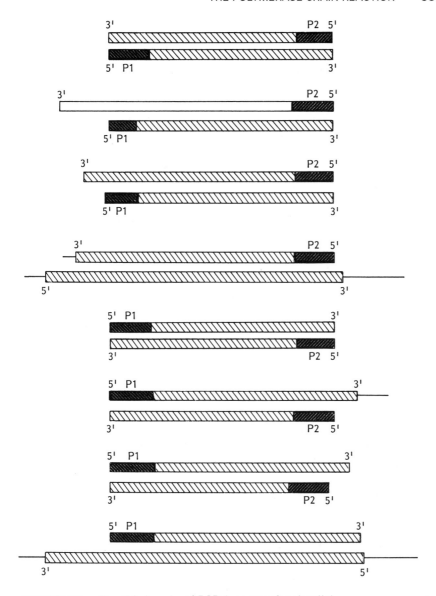

FIGURE 8.7: *The third cycle of PCR (see text for details).*

be performed at high temperatures which prevent the occurrence of nonspecific extensions. The most commonly used thermostable poly-merase is *Taq* polymerase. *Taq* polymerase is isolated from the organ-ism *Thermus aquaticus* and is available commercially from several leading biotechnology supply companies. Indeed the customer is spoilt for choice. Other thermostable polymerases available on the market include Vent polymerase which is isolated from *Thermococcus litoralis;* its use results in more accurate copying of the template than can be obtained by *Taq* polymerase. In addition, there is *Tth* polymerase which is isolated from *Thermus thermophilus*. The thermostable polymerase is the most expensive component of the PCR and is typically used at a concentration of one unit per reaction. This is sufficient to mediate amplification for up to 40 cycles in a PCR machine.

8.3.2 Target DNA (template)

One of the features of the PCR that makes it attractive as an alternative to gene cloning is its sensitivity. Amplification of single target molecules can be performed. This minimal requirement for template contrasts with gene cloning techniques and has been the primary reason for the expansion in the number of applications of PCR technology. However this high level of sensitivity is a major source of false positives caused by contaminating DNA. This is fully discussed in another book in this series [3].

Another advantageous feature in dealing with target DNA is that in some cases relatively impure samples can be used in the PCR. The lack of a requirement for purity has led to PCR being attempted successfully on blood spots, archival material, ancient DNA and even autoclaved bacteria. However, in many other cases elaborate sample preparation procedures must be followed to ensure amplification, as inhibitors such as heparin, EDTA or humic acids may be present in the sample.

The sequence of the target DNA can also affect the PCR amplification. If the DNA is extremely GC-rich (see Section 8.2.3) there will be difficulties in separating the DNA strands at high temperature. In such cases the addition of formamide or dimethyl sulfoxide (DMSO) can circumvent the problem. The concentration of target DNA in a PCR will influence the degree of amplification. Although single targets have been amplified, not every laboratory can achieve this level of sensitivity. Therefore, it is useful when setting up a PCR to dilute a target DNA to check the level of detection in a given system. An overabundance of target DNA can have a detrimental effect on amplification. Consider the first cycle (see *Figure 8.4*) of a PCR. If too high an amount of target DNA is present there will be a tendency to produce large quantities of primer-extended products. Primer limitation, and carry over of trace inhibitors and unfavorable reannealing kinetics may all contribute to poor amplification.

8.3.3 Primers

The primers are the components of the PCR that require most thought when designing experiments; therefore they will receive the most attention. Primers are chemically synthesized on a solid support in an oligonucleotide synthesizer. These machines are highly automated, making the task of primer synthesis an easy one. The primers used in a PCR will largely determine the success of the reaction. There are many rules to guide primer design and indeed there even exists software for choosing an optimum primer pair for the amplification of a target DNA of known sequence.

First there is an optimum size of between 18 and 25 nt for a primer. This size range ensures specificity of binding at the optimized temperature and keeps the cost of the primer to a minimum (the cost is approximately $3/£2 per nucleotide). Primers of greater length can be used if one wishes to increase the temperature of annealing of a primer to its complementary sequence. This is sometimes desirable in the event of spurious bands appearing on an agarose gel after analysis of the completed PCR mixture.

The choice of primer sequence is also of paramount importance. There are a number of important rules which, when applied, will increase the success rate of PCRs.

(i)   It is important to avoid inter- and intra-molecular complementarity when designing primers. In other words, the primers should not have sequences that allow them to hybridize to themselves and to each other. Such complementarity can lead to the formation of primer dimers which will sequester the components of the PCR and inhibit the formation of the desired PCR fragment.

(ii) It is also desirable to avoid having a high GC content or indeed a series of Gs and Cs in the primers. The primary reason for this rule can be appreciated if we return to the chemistry of nucleotide base pairing. The G:C bond is much stronger than the A:T bond as it has three rather than two hydrogen bonds involved in pairing (see Chapter 1). Hence a GC tract is significantly more difficult to melt than an AT tract. In the context of PCR, this phenomenon gives rise to the unfortunate situation of primer association even after heating to 95°C and, as a result, primer dimer formation and poor amplification of the desired product. This problem can be diminished by the addition of certain chemicals, such as formamide and DMSO, which lower the temperature required for melting out.

(iii) It is useful to estimate the optimum temperature for annealing of the primers used in a PCR. This value is termed $T_m-5°C$ which simply is the estimated melting temperature of the primer minus 5°C. A simple calculation can be used when determining this value.

$$T_m-5°C = 2(A+T) + 4(G+C)$$

Note the G+Cs have a greater temperature loading than the A+Ts. It is important to stress that this value is a rule of thumb and applies to primers of 16–22 nt long. The use of the calculation is less tenable outside this size range.

8.3.4 Magnesium chloride concentration

The concentration of $MgCl_2$ in a PCR is critical to its success. The reason for this is simple. If $MgCl_2$ were absent no amplification would occur. When one considers the interaction of $MgCl_2$ with the various components of the PCR its critical role becomes obvious. $MgCl_2$ is required for DNA polymerase activity; it influences annealing of DNA and primers, and it has a stoichiometric relationship with nucleotide triphosphates. This list is designed to illustrate the pivotal role of $MgCl_2$ in the PCR. To attempt to optimize the concentration of $MgCl_2$ in a PCR based solely on theoretical considerations, though interesting, would be time consuming if not indeed folly. A much simpler approach is to titrate the $MgCl_2$. This is done by setting up reactions with varying concentrations of $MgCl_2$ from 1 mM to 5 mM. In this way a $MgCl_2$ concentration optimum is established that suits the sample of interest.

8.4 PCR thermocycling

Having considered the principles and reaction components of the PCR it is time to discuss the mechanics. PCR amplification is achieved by repeated cycles of heating (denaturation), cooling (annealing) and extension. A typical cycle for amplification of a DNA fragment of 500 bp would be 95°C (denaturation) for 60 sec, 50°C (annealing) for 60 sec and 72°C for 60 sec. These representative values are given to allow the reader to appreciate what a thermocycling program is like but of course each PCR is designed and optimized individually. Thermocycling conditions are another area of optimization and depend on a number of factors.

(i)   The size of fragment to be amplified. Larger amplification products may need longer denaturation and extension periods. By the same token shorter fragments require less time for extension.

(ii)   The sequence of the primers and ionic conditions will determine what annealing temperature is best suited for the cycle. Therefore, depending on the experiment being performed, the annealing temperature may vary from 37° to 65°C.

(iii) The speed required in performing the PCR. Some experimenters prefer to use a two-step cycle rather than the standard three-step cycle. This simply involves heating and cooling steps. Polymerization occurs as the heating step proceeds. This variation works best with shorter fragments.

8.5 Conclusions

In the last decade the PCR has revolutionized scientists' ability to study gene structure and function. It has been established in molecular biology laboratories as a technical short cut to obtaining large amounts of DNA for subsequent studies. However, the most notable result of this new technology has been its applicability to different areas of the biological and medical sciences. Scientists, who have hitherto shied away from the use of genetic engineering techniques to gain insights in their respective fields, are embracing PCR technology. Recently, some notable examples have come to the public eye. The use of PCR in forensics, that is 'DNA fingerprinting' on dried blood stains, has been highly publicized in the media and indeed this technique is likely to influence the outcome in trials in the future. In the diagnosis of human genetic diseases and infectious diseases caused by microorganisms PCR is also beginning to play a crucial role. PCR can be used as a predictive test for inherited disorders and is now being used in the controversial area of prenatal diagnosis. In the case of infectious diseases, PCR is now used in the rapid diagnosis of infections such as human immunodeficiency virus (HIV), tuberculosis and a variety of venereal diseases. This allows clinicians to identify quickly the presence of an infectious agent and thus recommend a suitable regimen of treatment. PCR also allows the study of ancient DNA and archival material. In addition, scientists can now even study microorganisms that cannot be cultured using classical microbiological techniques, thus opening up a whole new source of information for the study of microbial evolution. The power of PCR and other amplification techniques is unquestionable as they have changed how scientists look at biological problems. In addition, these techniques have broadened the use of 'gene technology' to impact significantly on many scientific disciplines previously untouched by it.

References

1.   Mullis, K. and Faloona, F. (1987) *Meth. Enzymol.,* **155,** 335.
2.   Williams, J., Ceccarelli, A. and Spurr, N. (1993) *Genetic Engineering.* BIOS Scientific Publishers, Oxford.
3.   Newton, C.R. and Graham, A. (1994) *PCR.* BIOS Scientific Publishers, Oxford.

9 Final Scientific and Nonscientific Considerations

This book is designed principally for the newcomer to the area of genetic engineering. It has taken the reader from the precarious stages when genetic engineering came together as a result of the desire to isolate a particular piece of DNA. To understand the scope of the progress which has been made, it might be useful to recall those first faltering steps made in the direction of isolating genes.

One major difference for those starting out as genetic engineers now as compared with those who pioneered genetic engineering techniques is the relative ease with which experiments can now be performed. The ease and frequency with which one can isolate genes of interest comes in part from the methodological developments which have taken place over the years. Psychological barriers have also been removed. Those who were working in laboratories at the time before the first eukaryotic genes were isolated well remember the discussions which took place about the possibility of identifying one micro-organism in a collection of a million. The demonstration that it was possible made it easier for others to undertake such experiments without hesitation. The neophyte recombinant DNA scientists 20 years ago also faced the difficulty of gaps in their training. For instance, one could have been a biochemist with a broad knowledge of enzymes but little experience of micro-organisms. There was a tendency towards specialization. When genetic engineering came into being it was necessary to assimilate all of the diverse sectors and break down the barriers between them, which were sometimes intellectual, sometimes due to inadequate teaching materials and sometimes philosophical.

Many departments had become isolationist and had little contact with colleagues in other science departments. Suddenly there was mutual dependence. The final major difficulty for the genetic engineer at the start of this adventure was a lack of a ready supply of materials which were required for the successful completion of experiments. Enzymes which were essential for many of the steps involved in genetic engineering were not available from a catalog as is the case now. Time and effort had to be put into the generation of primary material for these experiments. The modern molecular biologist/genetic engineer can walk from stone to stone across a complicated river of technological steps by the simple device of buying kits from a variety of different manufacturers. Today, a major service industry has grown around the provision of the reagents required by molecular biologists using genetic engineering techniques. This was not always the case and those who were active in the early stages of genetic engineering now marvel at the ease with which DNA libraries can be obtained. Today it is possible to pay a few hundred dollars to obtain the gene library required. This has greatly accelerated the use of genetic engineering for the analysis of specific genes.

9.1 Words and messages

In the course of the development of the new era of genetic engineering a whole new lexicon has been put together (see Appendix A). A collection of bacteria that have successfully taken up vectors with foreign DNA is called a *library* or *bank*. The task of isolating a clone of interest ultimately requires the identification of one organism (containing the sequence of interest) from the bank. As the organism is grown up from a single source the task has become known as the *cloning* of DNA. The mixture of vector DNA and target DNA for cloning is called a *chimera,* and the process of putting those two fragments together appeared to be akin to recombination. Thus, the whole process has become known as *recombinant DNA technology*. An alternative way of describing the technology is *genetic engineering* and, as it implies, it involves moving around and modifying genetic material at will. The words which are used for all of these processes have, over the years, given rise to confusion and indeed antagonism. Cloning has been confused with a science fiction procedure by which individuals try to replicate themselves in an unlimited manner. Chimeras are also the name given to hybrid animals such as a goat and a sheep crossed at embryo level.

The use of the the term genetic engineer invites the accusation that one is 'playing God'. The ability to move DNA across genera, species and family boundaries is also one which has potentially worrying implications. This combination of words, many of which are emotionally loaded, inevitably provokes a response from society. The questions which arise are of two general types: ethical and concerning safety. The most obvious contexts in which concerns have been voiced to date are those of transgenic organisms and human gene therapy. The deliberate alteration of genotype and phenotype of organisms needs to be carefully monitored. With animals, the physiological impact of a transgene can be negative. In other cases it may be neutral and transgenic animals could be the preferred source for some therapeutic products which are of enormous benefit to society. This can be achieved without any apparent impact on the animal and by targeting the expression of the genes at sites such as the mammary glands. Society and regulatory bodies, and in particular scientists, must decide what type of genetic manipulation should be permitted.

The generation of transgenic plants may seem like a less emotive issue but the consequences of spread of a new trait in the wild is a concern, particularly in terms of the insertions that create resistance to pests and weedkillers. At the ethical rather than the safety level concern has also been expressed by consumer groups that they ought to be informed if they are buying genetically engineered products. The validity of these concerns needs to be established.

The ethical concerns with respect to animals and plants should be seen as a positive response to ensure that the benefits from the developments are not sullied by the generation of some inappropriately modified organism. A second ethical concern arises from the analytical possibilities afforded by the new technology. The number of diseases with genetic components which can now be rapidly and accurately defined by the use of DNA-based methods is ever increasing. The information obtained is of importance clinically and indeed can be of great use to a patient where a change of lifestyle can help avoid onset of disease. However, when this knowledge is allied to the fact that the diagnosis can be carried out at a very early stage following conception, the possibility of selection and counter selection for traits which may not be of clinical relevance (the sex of a baby for example) gives rise to concern. The methods developed to carry out these tests are at the very core of genetic engineering technology. The question is how can the technology be used in a way which will give benefits to society without the disadvantages? Genetic engineering is no different to any other technology in that it carries both positive and negative elements.

The other major question which arose with the advent of genetic engineering concerned safety. Once scientists realized that they had in their grasp the combination of methods, skills and material that would allow them to isolate human genes, it became clear that this methodology required careful consideration of the risk factors associated with it. In the vast majority of cases the risks were hypothetical. The initial reaction of the scientific community when the blueprint for genetic engineering became available was to hold a moratorium on the experiments for 1 year. This mature and voluntary move was ultimately interpreted as a signal that scientists really knew they had something very dangerous on their hands.

Since then controversy has always accompanied recombinant DNA technology. The possibility of generating dangerous and novel organisms which could be a threat was thought to be inherent in the technology. The extrapolation led to a general condemnation of genetic engineering which was and is unnecessary. The experience of many thousands of scientists carrying out multiple experiments since then serves to show that the methodology is in fact safe. The benefits of the technology have also been enormous and it now underpins many activities in the pharmaceutical industry. Recombinant DNA technology has also had a major positive impact on diagnostics both as the source of DNA probes and increasingly as the source of antigens for use in immunodiagnostic tests. The list is expansive and expanding. There can be few examples where the efforts of good science and such a diverse range of expertise could have been brought together into such a successful mosaic.

These achievements should not invite complacency: we should constantly review the work we are carrying out to ensure it is beneficial to mankind and not deleterious to our society and environment. As scientists, we must continue to build on the very solid base of knowledge which has been developed over the last 20 years, while constantly avoiding complacency in terms of ethical, social and legal aspects.

Appendix A

Glossary

Annealing or hybridization: the process whereby two initially separate, but complementary, nucleic acid molecules form a double-stranded structure.

Antibiotic resistance markers: gene coding for an enzyme capable of inactivating a particular antibiotic when it is expressed, used to select cells containing a particular vector.

Anti-sense RNA: this term normally refers to the RNA copy of a specific mRNA. Anti-sense RNA is used *in vitro* as a probe for hybridization and *in vivo* as a way of blocking gene transcription.

Autonomously replicating systems (ars): segments of DNA that, when inserted into a circular plasmid, confer the ability to replicate extra-chromosomally in yeast.

Background: a problem in most cloning procedures, where some (or in worst cases all) of the bacterial colonies which grow up do not contain vector molecules.

Cap: a methylated guanine residue which is added to the 5' end of eukaryotic mRNAs in a post-transcriptional reaction.

Cap site: the start site of transcription of a eukaryotic gene.

cDNA clone: a recombinant molecule containing the double-stranded DNA copy of a mRNA sequence.

Chimera: an embryo produced by introducing pluripotential cells into a normal embryo.

Chromosome walking: sequential cloning steps designed to bridge the gap between a probe that is to hand and a linked, but not immediately adjacent, region of a chromosome.

Clone: cells all of which contain the same DNA sequences.

Cohesive ends: overhanging ends of a double-stranded DNA molecule that are capable of hybridizing with complementary ends, also known as sticky ends.

Cosmid: a plasmid vector that contains *cos* sites and which can therefore be packaged into pseudo-viral particles.

Cos **sites:** the mutually adhesive termini of bacteriophage lambda.

Co-transformation: a procedure whereby two different DNA molecules, only one of which need contain a selectable marker, are mixed together and introduced into the genome of eukaryotic cells.

Cross-hybridization: the annealing of two nucleic acid molecules which are not perfectly complementary.

Deletion series: a set of clones, all derived from the same initial recombinant, but in which the insert lacks sequences at one of its ends because of treatment with an exonuclease.

Denaturation: a process whereby the two strands of a double-stranded nucleic acid molecule come apart as a result of heating or exposure to alkali conditions.

Differential screening: a method of screening cDNA libraries whereby two probes, differing in one or just a few sequences, are used as probes in two parallel *in-situ* hybridizations using duplicate lifts of the library.

DNA polymerase: an enzyme which copies a DNA or an RNA molecule to produce a DNA copy.

Electroporation: a method which uses an electrical pulse to introduce DNA into cells for transformation.

Endonuclease: a nuclease which cuts a nucleic acid molecule by cleaving between two internal residues.

Enhancers: eukaryotic gene regulatory elements that can confer tissue specificity of expression and which can activate gene transcription even when situated great distances away from the cap site of the gene.

Exon: a part of a nuclear RNA precursor which is joined together with other exons within the same RNA by splicing. The product of this reaction, the mRNA, is then exported to the cytoplasm.

Exonuclease: a DNA exonuclease which degrades a double-stranded DNA molecule by progressively removing nucleotides from its two ends.

Expression screening: a method of screening for a specific cDNA clone where the cDNA is inserted next to a promoter active in the host cell and an immunoassay or bioassay is used to detect the required clone.

Expression vectors: vectors capable of synthesizing proteins from inserted DNA sequences.

Fluorescent *in-situ* hybridization (FISH): a technique whereby a probe that can be detected using a fluorescently labeled reagent is annealed to chromosomes with the aim of localizing the target sequences.

Gene bank or library: a collection of recombinant DNA clones prepared by inserting DNA sequences into a vector. The number of clones required for a complete library depends on the size of the genome.

Gene fusion: a DNA segment containing parts of different genes, e.g. the promoter of one gene and the coding region of another gene.

Genome: the sum of the genetic information necessary to specify the formation of a living organism or a virus.

Genomic clone: a recombinant molecule containing genomic DNA, normally the term refers to a clone containing a gene and a variable amount of flanking DNA depending upon the method of cloning used.

Germ-line chimera or germ-line mosaic: a transgenic mouse in which some or all of the germ cells contain the transgene and which is therefore able to pass the gene on to its progeny.

Insert: a target for isolation by DNA cloning, e.g. the cDNA copy of an mRNA.

Intron: part of a nuclear RNA precursor which is removed and degraded within the nucleus to yield mRNA.

***In-vitro* packaging:** a method whereby DNA flanked by phage packaging signals is encapsidated *in vitro*.

Lift: a replica copy of a bacteriological plate bearing bacterial colonies or phage plaques made by touching an inert filter on to the plates surface.

Ligases: enzymes that catalyze the joining of DNA strands.

Maxam and Gilbert (chemical degradation) procedure: a DNA sequence analysis method.

MCS (multi-cloning site or polylinker): a short DNA sequence, found in most vectors in common use, which contains many closely spaced restriction enzyme cleavage sites.

Melting temperature or T_m: the temperature at which the two strands of a double-stranded nucleic acid molecule come apart.

Messenger RNA (mRNA): an RNA molecule that contains the genetic information necessary to encode a protein.

Microsatellites: a class of highly variable markers in human DNA which are composed of di-, tri- and tetra-nucleotide repeats, for example $(CA)_n$ or $(CCA)_n$.

Mini-preparations: DNA prepared from small bacterial cultures derived from individual colonies in a cloning experiment.

Mosaic: a transgenic animal in which the transgene is present in only a fraction of cells within the animal.

5' and 3' noncoding regions: untranslated regions at the 5' and 3' ends of mRNA sequences.

Northern blotting: a process whereby RNA molecules are separated by gel electrophoresis, transferred to a filter and hybridized with a specific probe with the aim of detecting complementary target molecules.

Operons: closely spaced bacterial genes which function in a common metabolic pathway.

Phenotype: characteristics of a cell or organism that are defined by the expression of genes within the cells.

Plaque: an area of a bacteriological plate where the bacteria are dead, or grow slowly because of infection by a virus.

Plasmid: extrachromosomal DNA element capable of autonomous replication.

Poly(A) tail: a tract of A residues of approximately 100–200 nucleotides in length that is added enzymatically to the 3' end of a mRNA.

Polymerase chain reaction (PCR): a method of specifically copying, and amplifying, a part of a nucleic acid chain.

Pooled (amplified) libraries: for long-term storage phage particles or bacterial colonies on bacteriological plates are eluted in a mixed pool, containing representatives of every different recombinant on the plate.

Primer: a short nucleic acid molecule which, when annealed to a complementary template strand, provides a 3' terminus suitable for copying by a DNA polymerase.

Primer extension: a method of establishing the start site of transcription of a gene.

Probe: a nucleic acid sequence that is complementary to part or all of a target which is to be detected by hybridization. It is usually 'tagged' by the incorporation either of radioactively labeled nucleotides or of nucleotides which are chemically modified in such a way that they can be identified immunologically.

Pulse field gel electrophoresis (PFGE): a form of gel electrophoresis that allows extremely long DNA molecules to be separated from one another.

Recombinant DNA: a molecule of DNA generated by the insertion of foreign DNA into a vector.

Replication origin: a segment of DNA that acts as the start site of DNA replication.

Restriction enzymes: enzymes which cleave double-stranded DNA into discrete pieces by cleaving at defined recognition sequences.

Restriction fragment-length polymorphism (RFLP): a localized difference in the genome structure of individuals within a population that produces a difference in their restriction maps.

Restriction map: a physical map of a piece of DNA showing the position of cleavage of one or more restriction enzymes.

Ribosomal RNA (rRNA): an RNA molecule that forms part of the ribosome, the bipartite structure where mRNAs are translated into proteins.

RNA polymerase: an enzyme which copies a DNA or an RNA molecule to produce its RNA copy.

Sanger (dideoxy or chain termination) procedure: the standard method of DNA sequence analysis.

Screening: the process of identifying the clone containing the desired DNA sequence.

Shuttle vector: a vector that is able to transform both *E. coli* and some other organism; constructions and mutations can be made in *E. coli* and the effects studied in the alternate host.

S1 mapping: a method of determining the transcriptional organization of a gene.

Southern blotting: a process whereby DNA molecules are separated by gel electrophoresis, transferred to a filter and hybridized with a specific probe with the aim of detecting complementary target molecules.

Splicing: the series of steps within the nucleus whereby introns are removed from the nuclear precursor to form a mRNA and the exons are joined together.

Sticky ends: overhanging ends of a double-stranded DNA molecule that are capable of hybridizing with complementary ends. Also known as cohesive ends.

Stringency of hybridization: the combination of temperature, salt and formamide concentration which determine the degree of cross-hybridization that can occur in an annealing reaction or during post-hybridization washing.

Subcloning: the process wherein a purified DNA molecule is inserted into a vector and isolated by gene cloning.

Subtraction hybridization (subtraction enrichment): a method of screening cDNA libraries whereby two probes, differing in one or just a few sequences, are hybridized together so that the sequences common to the two populations anneal. The nonannealed sequences are then used to prepare a cDNA library and/or a probe for screening.

Target: a nucleic acid molecule which is to be detected in a hybridization reaction or which is to be isolated by molecular cloning.

Transcription: the process whereby a DNA molecule is copied into RNA.

Transcription factor: a protein that regulates or facilitates the transcription of a gene.

Transfer RNA (tRNA): the intermediaries which carry amino acids to the ribosome and which direct their position of insertion into the polypeptide chain.

Transformation (DNA transformation): the process whereby a DNA molecule is introduced into a living cell. The term is normally reserved for those cases where the DNA molecule has the capacity to be stably maintained in the host cell.

Transgenic organisms: animals or plants the phenotype of which have been modified by the insertion of foreign DNA into the cells of the organism.

Transient expression: a technique in which DNA is introduced into eukaryotic cells and its transcription is analyzed after it has reached the nucleus, but before it has integrated into the genome.

Translation: the process whereby rRNAs, and their associated proteins, decode the linear sequence of information contained within the mRNA to form a protein.

Vector: a DNA molecule that is capable of autonomous replication in a cell and with restriction sites able to accommodate additional DNA sequences.

Yeast artificial chromosomes (YACs): recombinant DNA molecules that contain very large DNA inserts and which replicate in yeast cells as linear, mini-chromosomes.

Appendix B

Further reading

Chapter 1

Avery, O.T., MacLeod, C.M. and McCarty, M. (1944) *J. Exp. Med.,* **79,** 137.

Cairns, J., Stent, G.S. and Watson, J.D. (1966) *Phage and the Origins of Molecular Biology.* Cold Spring Harbor Laboratory Press, Cold Spring Harbor, NY.

Griffith, F. (1928) *J. Hyg.,* **27,** 113.

Portugal, F.H. and Cohen, J.S. (1980) *A Century of DNA.* MIT Press, Cambridge, MA.

Watson, J.D. (1980) *The Double Helix: A Norton Critical Edition* (G.S. Stent, ed.). Norton, New York.

Williams, J., Ceccarelli, A. and Spurr, N. (1993) *Genetic Engineering,* BIOS Scientific Publishers, Oxford.

Chapter 6

Grunstein, M. and Hogness, D.S. (1975) *Proc. Natl Acad. Sci. USA,* **72,** 3961.

Maniatis, T., Fritsch, E.F. and Sambrook, J. (1989) *Molecular Cloning,* 2nd Edn. Cold Spring Harbor Laboratory Press, Cold Spring Harbor, NY.

Chapter 8

Barry, T., Colleran, G.J., Glennon, M., Dunican, L.K. and Gannon, F. (1991) *PCR Meth. Appl.,* **1,** 51.

Innis, M.A., Gelsand, D.H., Sninsky, J.J. and White. T.J. (1990) *PCR Protocols.* Academic Press, San Diego.

McPherson, M.J., Quirke, P. and Taylor, G.R. (1992) *PCR: A Practical Approach.* IRL Press, Oxford.

Newton, C.R. and Graham, A. (1994) *PCR.* BIOS Scientific Publishers, Oxford.

Index